Educational Research in Higher Education: Methods and Experiences

RIVER PUBLISHERS SERIES IN INNOVATION AND CHANGE IN EDUCATION - CROSS-CULTURAL PERSPECTIVE

Volume 12

Nowadays, educational institutions are being challenged when professional competences and expertise become progressively more complex. This is mainly because problems are more technology-bounded, unstable and ill-defined with the involvement of various integrated issues. To solve these problems, it requires interdisciplinary knowledge, collaboration skills, innovative thinking among other competences. In order to facilitate students with the competences expected in professions, educational institutions worldwide are implementing innovations and changes in many aspects.

This book series includes a list of research projects that document innovation and change in education. The topics range from organizational change, curriculum design and innovation, pedagogy development, to the role of teaching staff in the change process, students' performance in the aspects of not only academic scores, but also learning processes and skills development such as problem solving creativity, communication, and quality issues, among others. An inter- or cross-cultural perspective is studied in this book series that includes three layers. First, research contexts in these books include different countries/regions with various educational traditions, systems and societal backgrounds in a global context. Second, the impact of professional and institutional cultures such as language, engineering, medicine and health, and teachers' education are also taken into consideration in these research projects. Thirdly, individual beliefs, perceptions, identity development and skills development in the learning processes, and inter-personal interaction and communication within the cultural contexts in the first two layers.

We strongly encourage you as an expert within this field to contribute with your research and make an international awareness of this scientific subject.

For a list of other books in this series, www.riverpublishers.com

Educational Research in Higher Education: Methods and Experiences

Editor

José Gómez Galán

Metropolitan University, PR, USA and
University of Extremadura, Spain

LONDON AND NEW YORK

Published 2016 by River Publishers
River Publishers
Alsbjergvej 10, 9260 Gistrup, Denmark
www.riverpublishers.com

Distributed exclusively by Routledge
4 Park Square, Milton Park, Abingdon, Oxon OX14 4RN
605 Third Avenue, New York, NY 10158

First published in paperback 2024

Educational Research in Higher Education: Methods and Experiences / by
José Gómez Galán.

Routledge is an imprint of the Taylor & Francis Group, an informa business

Publisher's Note
The publisher has gone to great lengths to ensure the quality of this reprint but
points out that some imperfections in the original copies may be apparent.

While every effort is made to provide dependable information, the
publisher, authors, and editors cannot be held responsible for any errors
or omissions.

ISBN: 978-87-93379-66-4 (hbk)
ISBN: 978-87-7004-457-8 (pbk)
ISBN: 978-1-003-33803-1 (ebk)

DOI: 10.1201/9781003338031

Contents

PART I: Research Methods in Higher Education: Theoretical Framework

1 Quantitative Research   **3**

Omar A. Ponce and Nellie Pagan

2 Qualitative Research   **45**

Omar A. Ponce and Nellie Pagan

**PART II: Models and Experiences in Higher
Educational Research**

6 Measurement Adult College Student Experience: A Phenomenological Analysis   **139**

Zobeida González-Raimundí

7 Equality Studies in Higher Education: A Model of Qualitative Research Method   **145**

Mariwilda Padilla Díaz and José Gómez Galán

8 Climate Change Education: A Theoretical Model   **161**

José Gómez Galán

Preface
Innovative Educational Research Models
for the Improvement of Higher Education

"Nothing has such power to broaden the mind as the ability to investigate systematically and truly all that comes under thy observation in life"
Marcus Aurelius

"The joy of seeing and understanding is the most perfect gift of nature"
Albert Einstein

"Somewhere, something incredible is waiting to be known"
Carl Sagan

The work presented here is an effort by multiple authors to report not only the current state of affairs in the complex field of educational research in higher education, but a commitment to new research models that better meet today's needs and for which answers must be provided.

Although the main objective of the work could be understood as being a manual accessible to both early stage researchers – for example those who are preparing their doctoral thesis – and veterans who wish to keep abreast of what is being done at present in educational research, it is in no way limited to this. It is also a work of reflection that aims to inspire readers to create their own research models and to introduce them to the wealth and possibilities currently available to the world of research, thanks to new computer and telematic tools which are opening up a new research paradigm for us.

From a methodological point of view, there is still much work to be done in the field of educational research. Placed as it is within the main problems that affect research in the social sciences, and in many cases vilified by theorists of experimental sciences, research in the field of education gains predominance because of the fundamental challenges it faces today.

This is firstly because we are talking about a scientific field, such as education, which not only contributes decisively in building the present world, but it is essential when looking towards the future. Everything that we are currently applying in this context is not something that we will observe immediately; instead we will reap the benefits in the next generation which is currently being educated. Therefore, in education it is also necessary to know everything that affects the world today and its multiple problems, in a multidisciplinary context, as education should primarily be the engine that contributes to the improvement of society and the evolution of human beings. Without a doubt, it must be the main instrument that responds to the most important problems that need to be solved.

This implies, therefore, that educational research should be investigated globally. It should not only focus on teaching and learning processes, as so often is the case, but be an investigation that is capable of responding to the needs in multiple fields of knowledge that are directly or indirectly related to our science.

Secondly, we are at a historical turning point. The emergence of information and communication technologies (ICT) is bringing about a complete transformation of all aspects of our lives. Whether we are in the last stage of the industrial revolution, an "informational revolution" as advocated by Castells, or a new evolutionary period in the evolution of civilization, the fact is that we are entering into a new era in which the digitisation of information and communication processes will shape our existence. In this era, science and scientific research cannot stand in the side-lines as they will be the main beneficiaries of these changes.

Even though we do not know what will happen tomorrow, as is the case in every revolutionary moment in the history of mankind, we can be absolutely sure that it will be different from what we have today. Using Bauman's words we are living "liquid times". Times when you need to seek guidance and structures on which we are based. Despite the great potential that ICTs offer, maximising their benefits and minimising their drawbacks is of most use. These clearly exist extensively and the field of educational research is no different from other fields. The great potential that ICTs open up in that huge volumes of information can be managed must not make us forget that our working material is essentially human beings, society, and in general, the environment in which we find ourselves, with all the forms of life it contains. It makes little sense to carry out highly complex studies, statistical studies for example, through quantitative methodologies that would rival those used in the experimental sciences, if we miss the point on what the questions and needs we have today in education are, which should be what guides our work.

This should all be applicable in the field of higher education. It is at this level where highly qualified professionals are prepared to respond to the major challenges of society, and also where humans who should contribute to building a better world are educated. Indeed, we are far from understanding that the academic world is only a place of professional training. Nevertheless, it is still, without a doubt, a major educational area aimed at human growth.

Certainly, the crisis of the positivist paradigm has led to a critical reflection on the characteristics required for educational research in higher education, in order to respond to real world needs. We are placed in a context where the number of interdependent variables is so high that on many occasions the use of models from the field of experimental science is insufficient to respond to the essential knowledge of the main problems.

The most useful approach is most likely to start from methodologies that are closer to the complexity of humans, specific to the field of social sciences and humanities. From this starting point, methods and techniques of a quantitative nature specific to experimental sciences should be applied when required, thereby merging paradigms but always serving the objective pursued. The method cannot be above the goals pursued, or above the questions that we face or the questions that we ask ourselves to find out what the essence of reality is. Rabindranath Tagore said that asking questions is the evidence that one is thinking. To which we would add that there is nothing more specific to research than thinking: it is a human feature and is what should become the basis of the investigation. As research is done to answer questions it is necessary to give educational research a human focus and inspire it to investigate what is needed.

This book will help in both knowing how to investigate and how to make research more human. Although the most comprehensive statistical methodologies can be used, they must always be justified and ready to serve the objectives of what we truly need. These objectives must not be at the service of the method. Rigor is essential of course, but for something meaningful. Saturdays were made for man, not man for Saturdays. Quantitative, qualitative and mixed methodologies, positivist and naturalist paradigms, experimental and quasi-experimental designs, statistics, active research, software, etc., are common words in today's language of educational research. This work is intended to respond to the complexity of research in higher education, and to clarify the main features of it in an accessible way. It will not only focus on presenting a set of models and methodologies to support the researcher nor will it propose any specific approach. It must not be forgotten that we are frequently influenced by trends, by methods that are successful in a particular field and that are generalised to a wider application without being accompanied by a rational reflection to justify them. Conversely, techniques that were traditionally successful and offered excellent results are vilified, cast aside and forgotten about. Each researcher must create their own models or use those that are actually useful for their goals: innovative models that contribute to improving the academic world.

Most manuals on educational research in higher education are usually a description of the various quantitative, qualitative and mixed approaches that have been used in recent decades. It remains true that a very wide range of these are needed in the current context of a global and complex society, made up of multicultural elements intimately connected with various economic, social, political and, of course, educational dimensions, which leads to facing research processes that vary significantly in nature. Nevertheless, the method will always be insufficient against such a volume of variables, so this element should not be the backbone of the concept of educational research.

A much more comprehensive view of the problems in higher education is required, which arise from the new digital society and globalised world in which we live. Therefore, this work is above all an invitation for the reader to reflect upon their own research processes and to build them, offering the most appropriate means to enable this. This is done in two parts. In the first part the latest theories of research in this field of study are presented; the main paradigms, methods and techniques and the reasons thereof. In the second part, examples of different research designs applied in practice in the field of higher education are presented. Within this work varied models applied to areas such as teaching methods in the classroom, educational innovation, equality studies, educational technology, etc. can be found, which are all up-to-date and capable of absorbing the main challenges and demands we face today. This book will hopefully fulfil the functions and achieve the goals that we have pursued: to simply be useful to the educational researcher in higher education.

José Gómez Galán

List of Figures

List of Tables

List of Abbreviations

ATI	Aptitude Treatment Interaction
CCE	Climate Change Education
EEV	Education in Environmental Values
EE	Environmental Education
ECPI	Estatuto de la Corte Penal Internacional
FRA	European Union Agency for Fundamental Rights
GPA	Grade Point Average
ICT	Information and Communication Technologies
IQ	Intelligenz-Quotient
IPCC	Intergovernmental Panel on Climate Change
IET	Intraeducacional Educational Technology
IM	Intra-Medium
NRC	National Research Council
PAEG	Prueba de Acceso a Enseñanzas de Grado
R&D	Research and Development
TES	Supraeducacional Educational Technology
CEPES	The European Centre for Higher Education/Centre Européen pour l'Enseignement Supérieur
CEPAL	The Observatory for Gender Equality of Latin America and the placeCaribbean
SPSS	The Statistical Package for Social Science
UN	The United Nations
UNESCO	The United Nations Educational, Scientific and Cultural Organization
UNFPA	The United Nations Population Fund

PART I

Research Methods in Higher Education: Theoretical Framework

1

Quantitative Research

Omar A. Ponce and Nellie Pagan

Metropolitan University, PR, United States

1.1 Introduction

The quantitative research model is the first scientific research model that was used to study the problems of education. It was imported from natural sciences to the education field (Pauls, 2005). Research methods can be classified into experimental designs and non-experimental designs. Figure 1.1 summarizes these designs. The quantitative research model has the following distinctive methodology which emanate from the positivist and post-positivist philosophy on which it is based:

a. **Emphasis on the study of causal relationships.** The emphasis and strength of quantitative research is the study of causal relationships. A causal relationship means that the manifestation of a situation, event, condition or activity (independent variable) produces direct consequences or reactions, in a chain, in another situation, event, condition or activity (dependent variable). The most common example in education from a relationship of cause and effect is the teaching- learning process. Teachers teach to produce student learning. For a relationship to be considered causal, the "cause" must precede the effect. Quantitative research methods are based on the assumption that human behavior is influenced by the environment or by the external social reality to the individual. The social environment is seen as an independent entity to humans. An example of this would be the following positioning statement; a child who grows up in a hostile environment becomes an aggressive adult. Based on this premise, quantitative research focuses on identifying those universal cause and effect relationships that influence human behavior. For example, which environmental factors in schools (causes) affect student learning (outcomes)? The central argument of the

3

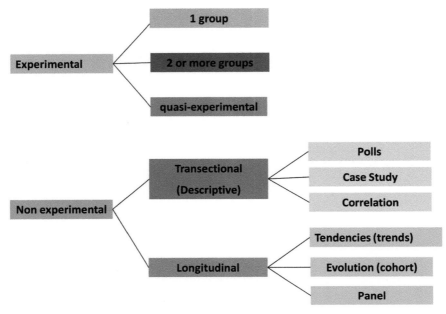

Figure 1.1   Quantitative research designs.

quantitative model is that if you can identify these universal factors in education (teaching techniques, curricula or more effective evaluation methods) that affect student learning, educators can anticipate results, prevent situations or modify events.

b. **Emphasis on measurement and numerical description of the phenomena studied.** Since it is assumed that the environment or social reality is external to humans, knowledge is identified in the characteristics or properties of social phenomena being studied. Therefore, the quantitative researcher uses measurement tools (questionnaires, scales, rubrics) and statistics to accurately measure the properties of social phenomena studied. Thus, the property studied can be described numerically or the magnitude of the effect determined on the student. In the field of education this possibility is very attractive to accurately determine the effectiveness of instructional techniques, curricula or academic programs.

c. **Emphasis on objective and systematic research methods.** Since it is assumed that the environment or educational reality and the individual entities are independent of each other, to study these relations of cause and effect quantitative researchers pay careful attention not interfere or interact with the natural manifestation of the phenomena they study or with the behavior of people who constitute the study participants. To do this, they carefully plan the research method. Thus, quantitative research argues that studies systematically and objectively the phenomena of education and its effects on students. The means to achieving this neutrality (objectivity) is through research designs and its methods.

d. **Emphasis on studies with random samples of participants to generalize findings.** The quantitative research model emphasizes the use of random sampling in selecting study participants in order to generalize the findings. As the premise is that educational and human behavior phenomena are independent entities, then it is argued that it is possible to study these causal relationships with random samples and this makes it easier to generalize the study findings. A random sample is a thorough and representative selection of study participants. The argument is that with representative samples of participants it is possible to study these causal relationships in less time and with less money. This is attractive since the possibility of a classroom study and of generalizing the findings to schools and education systems makes quantitative research an attractive tool for the development of education.

e. **Emphasis on statistical analysis as a tool for interpreting data.** The quantitative method uses statistics as a basis for the analysis and interpretation of data. It is argued that many of the environmental influences on student behavior are imperceptible to the human eye. Statistics are an objective tool to identify and accurately determine the probability of patterns occurrence and trends of causal relationships that are not caught by the human eye.

1.2 Experimental Research Design

Experimental research design is the first type of scientific research that was conducted in education. As research design, its distinctive is to test relationships of cause and effect, creating the treatment and conditions for this (experiment), and thus collect data. For example, what effect teaching technique X has (cause- treatment) in the learning of students (effect-result). Once the conditions are created for the experiment, the researcher manipulates a situation (independent variable, treatment or cause) to study its consequences in other areas (dependent variable or effect). This design has been used in education to establish the validity and effectiveness of teaching techniques, curricula or programs to promote learning. The logic of the experimental design is linear and simple. A cause and effect is identified. That relationship becomes operational in the form of a hypothesis or in the form of a tentative statement of the causal relationship it is expected to observe. For example, there is a relationship between the number of hours of watching television and academic achievement. With the experiment test the relationship to determine whether or not this is occurring. If the relationship occurs, the hypothesis is confirmed and if not, it is rejected. Figure 1.2 illustrates this logic.

For example, say that after several years teaching health to fifth graders, the team of teachers at this level is convinced that the best way to prevent students from experimenting with smoking is to make them aware of how damaging smoking is to health. Their hypothesis is that if students are instructed about the adverse effects of smoking on health, students lose the curiosity and interest. They decide to design a unit in the health curriculum to test the hypothesis. They select one of the fifth grade groups and for one semester expose them to lectures, films, visits to hospitals and interviews with smokers to know and experience firsthand the adverse effects of smoking (the cause, treatment or independent variable of the experiment) and observe if their hypothesis is not confirmed or (effect of experiment or the dependent variable). The experiment begins by measuring the knowledge and attitudes of students about smoking. They are exposed to the health unit. At the end of the semester the knowledge and students' attitudes about smoking is re-measured. If students change their attitudes at the end of the experiment, and perceive smoking as unhealthy, the teachers will have confirmed their hypothesis. Then they may conclude that they were right and that emphasizing the negative effects of smoking prevents fifth grade children to experiment with smoking. If the result of the experiment would have been the opposite, then teachers must reject their hypothesis and assert that there is no relationship between knowledge about the adverse effects of smoking on health and the tendency to experiment with smoking. In educational research there are three common types of experiments: the experiment with a group, the experiment with control groups and the quasi-experiment. Let's examine these designs separately.

a. **Experiments with a group** (One group pretest and post-test). The example used in the previous section illustrates this design. Figure 1.3 diagrams the experiment with a group that is evaluated before and after the study:

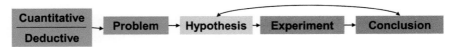

Figure 1.2   Structure of experimental design.

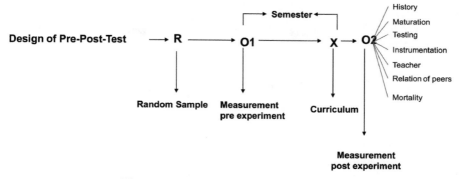

Figure 1.3 Experimental design – 1 group.

Selection of the study sample. The experiment begins with the random selection of students who will constitute the experimental group or the study sample. In Figure 1.3 they are illustrated with the letter R. A random sample means that every fifth grader in the school had the same opportunity to be selected to participate in the study.

Pre-experiment measurement. The group before undergoing the experiment is evaluated. For example, students' attitudes toward smoking are measured. Figure 1.3 illustrates this with the symbol 01. 01 means the group attitudes before the experiment. That measurement is the starting point or comparison study to determine whether there are changes in attitudes after the experiment.

Experiment. Selected students (study sample) are given a curriculum oriented to expose them to the harmful health effects of smoking. In Figure 1.3, the curriculum is illustrated by the symbol X. The curriculum is the treatment of the experiment or the independent variable that the researcher manipulates to observe the effect. The effect in this example, if they occur, will be changes in student attitudes toward smoking. The curriculum is the "independent variable" because it is the "treatment" that the researcher identified and designed to "manipulate in the experiment." Manipulation comes in the form of curriculum content, teaching strategies and the duration of the experiment. The experiment is commonly depicted in the literature with the following symbol: 01 X 02.

Post-experiment measurement. The experiment concludes with a measurement of students' attitudes toward smoking, after a semester or the end of the curriculum. In Figure 1.3 this is illustrated by the symbol 02.

Post-experiment measurement is compared with pre-experiment measurement to determine whether there were changes in attitudes and determine whether the study hypothesis is accepted or rejected. With this data (measurement), the quantitative researcher can assert whether there is a relationship between changing attitudes towards smoking and the act of smoking due to curriculum content, teaching strategies and the duration of the experiment.

Strengths of the experiment with a group and its implementation challenges. The strength of the experiment with a group is a real possibility to study educational relations of cause and effect. By focusing on the study of a group its implementation is made feasible in public and private education systems because it involves major changes to the organizational structure of the institution. For example, systems of primary and secondary education tend to operate with school organizations which place students in groups of similar grade point averages, talents or abilities. This is done in order to meet their educational needs. Creating random groups for an experiment may involve altering the school organization and create discomfort with parents and students. There are school principals and district superintendents that allow the development of experiments in their schools, but without altering the school organization and operation of the school. In the case of private schools, some operate with reduced tuition because they are limited to the physical size of their buildings. Others control the size of enrollment to provide individualized attention that parents paying for the education of their children prefer. Being able to create comparable groups is impossible in some private schools because their enrollments have one or two groups by academic grade.

Critique on experiment with a group. Criticism of the experiment with a single group is the possibility that the researcher can never be sure whether the effect measured by the experiment was not due to treatment or the independent variable. Returning to the example of the curriculum to change attitudes towards smoking, the criticism is that the researcher is not satisfied or can ever establish beyond doubt that the curriculum is what caused the attitude change because the only evidence generated by the experimental data are pre and post test. Given this critique, quantitative researchers use experiments with control groups for comparison points to explain the experimental data.

b. **Experiments with comparison groups.** In this design the experiment is carried out, but involves the use of two similar groups in order to compare and explain the findings. The logic design is to expose one of groups to the treatment and the other not. For example, say that a group is exposed to the curriculum to raise awareness about the adverse effects of smoking and the other did not receive any guidance. At the end of the experiment, if the curriculum is effective, the group receiving the treatment should reveal changes in attitudes and knowledge towards smoking. These changes are most evident when compared with the group receiving no treatment. Figure 1.4 charts this type of research design.

Sample selection. Students are randomly selected to form two groups. One group is known as the **experimental group** (Group 1 of the diagram) and the other is referred to as the **control group** (group 2 of the diagram). Both groups have the same number of members with similar characteristics. Similarities in characteristics could be age, height and weight, IQ intelligence, similar personalities, language skills, values and attitudes.

Pre-experiment measurement. Both groups are evaluated before undergoing the experiment. For example, students' attitudes toward violence are measured. In Figure 1.4 this is illustrated with the symbol 01 for the experimental group and 03 for the control group. 01 is the measurement the attitudes of the experimental group before the study. 03 is the measurement of the attitudes of the control group before the experiment. That measurement is the starting point or comparison groups to determine if there are changes in attitudes after the experiment.

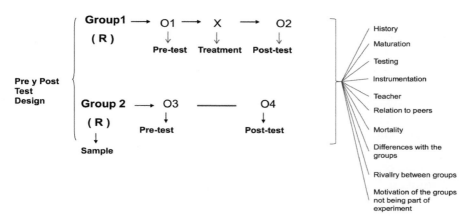

Figure 1.4 Experiment with control groups.

Experiment. Both groups are exposed to the experiment. Say you want to measure whether television violence causes the sixth graders see violence as acceptable. The experimental group will observe movies with violence and control group observes movies without violence. In Figure 1.4 this diagram is as follows: The experimental group is symbol X to denote the treatment or movies with violence and control group with the symbol __ to denote different treatment or movies without violence.

Post-experiment measurement. The experiment concludes with a measurement of student attitudes toward violence. In Figure 1.4 this is illustrated with a 02 for the experimental group and 04 for the control group. Post-experiment measurements are compared with pre-experiment measurements of both groups to determine if there were changes in the attitudes of students toward violence and determine whether to accept or rejected the study hypothesis. With this data (measurements), the quantitative researcher can assert whether there is a relationship between attitudes toward violence in both groups receiving different treatment. For example, if children exposed to violent programs show greater tolerance for this than the control group, the researcher can conclude that television violence caused attitudes.

Strength of the experiment with two groups and implementation challenges. This design allows the study of causal relationships in a real way from experimenting with two groups. Having a control group does help explain the changes in the group receiving the treatment, if the expected changes occur. The challenge of experimenting with comparison groups is its implementation when creating comparison groups by random sampling. As mentioned in the previous section, public schools work with school organization of groups consisting of skills and academic ability. Creating comparable groups involves altering the school organization. However, this design is feasible to implement in many private schools because they do not alter the organizational structure of the institution. For example, in small private schools, the trend is to have one and not more than two groups by academic grade. So rearranging the groups to make them comparable, rarely involves substantial changes in the operation of the institution.

Critique of experimenting with two groups. The biggest critique of experimenting with two groups is the possibility that the researcher will never be certain whether the effect measured by the experiment was due to treatment or the independent variable. Returning to the example of the curriculum to change attitudes towards violence, critique would be that the researcher can never establish beyond doubt that the curriculum is what caused the change in attitudes. Although the researcher does their best to control external variables of the study, and uses a control group; the only evidence of the experiment is data generated from pre and post tests. Given this critique, quantitative researchers have turned to experiments with three and four treatment groups to compare and explain the experimental data (see Figure 1.5). These designs use the same logic and have also been critiqued.

c. **Quasi-experimental research.** A quasi-experimental study follows the same format as the experimental design with two groups. Comparison groups are used to test a treatment, measure a relationship of a suspected cause and effect, but the sample is not random. This limits the generalization of the data or its external validity. In a quasi-experimental study, the researcher attempts to balance, if possible, their groups, to make them comparable, but recognizes that these are not equal. Figure 1.6 diagrams the quasi-experimental design:

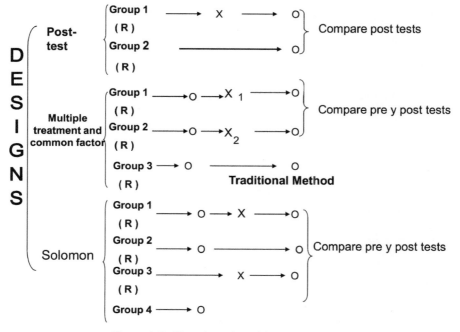

Figure 1.5   Experimenting with control groups.

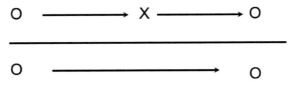

Figure 1.6   Quasi-experimental design.

Sample selection. Students are selected to form two groups. One group will be the **experimental group** and the other the **control group**. Sample selection is not random and that makes the study quasi-experimental. Despite this, the researcher strives that both groups have similar characteristics and are comparable. The similarities in characteristics could be the age, IQ intelligence, level of maturity, language skills, or values and attitudes.

Pre-experiment Measurement. Both groups are evaluated before undergoing the experiment. Figure 1.5 illustrates the symbol 01 for the experimental group and 03 for the control group. 01 means the measurement the experimental group have before the study. 03 means the measurement the control group have before the experiment. That measurement is the starting point or comparison of groups to determine if there are changes after the experiment.

Experiment. Both groups are exposed to the experiment. The experimental group received the treatment to be observed as the independent variable and the control group received other treatment. In Figure 1.5 this diagram is as follows: The experimental group with symbol X denotes the treatment and control group with the symbol __ denotes different treatment.

Post-experiment measurement. The experiment concludes with a measurement of the post-test groups. In Figure 1.5 this is illustrated with a 02 for the experimental group and 04 for the control group. Post-experiment measurements are compared with pre-experiment measurements of both groups to determine if there were changes in the experimental group and determine whether to accept or rejected the study's hypothesis. With this data (measurements), the quantitative researcher can assert whether or not there is a causal relationship, but recognizes that its findings cannot be generalized to other students than those who participated in the study.

Strengths of the quasi-experimental design. The biggest challenge of implementing experimental designs is to create experimental groups based on random samples. This limitation disappears with Quasi-experimental design. In this design, the researchers evaluate the existing groups in educational institutions and establish the initial scores prior the experiment. For example, say that the experiment is about a new technique for teaching of writing. The researcher measures the level of achievement in writing of their experimental groups before the study. This way they can determine how the groups were in writing prior to the experiment and use these scores as points to compare the groups after the experiment. This makes possible the implementation of the quasi-experimental design in any educational system because it is inserted into existing school reality without altering the organization or its operation. **Critique of the quasi-experimental design.** The biggest critique of quasi-experimental design is its sampling. Some researchers see it as an inferior design to the experiment by its limitation to generalization of the findings or external validity by not using random samples. My experience is that the quasi-experimental design is a real alternative to test causal relationships in educational settings of the country. As we will study in the following chapters, few designs provide this possibility. This is an undeniable strength of the quasi-experimental design.

1.3 Challenges to the Validity of the Experimental Design

In the field of education, the effectiveness of the experiment is questioned by the complexities of establishing beyond doubt, as in the example that was used at the beginning, whether it is the curriculum that is the cause of the changes in student's attitudes and not the effect of other "intervening variables" or "external variables." In Figure 1.3 these challenges are illustrated with the following terms: History, Maturation, Testing, Instruments, Teacher, Relation to peers and Mortality. For example, "intervening variables" to the study that can induce changes, regardless of the curriculum are:

a. **History.** If we use the example of curriculum to change attitudes towards smoking, this concept refers to the possibility that the attitudes of one or more of the students can change because of external events of the experiments. For example, a relative of one of the participants dies of lung cancer by smoking and this influences the attitudes of one or more of the children in the experiment. Another possibility is that during the months of the experiment a campaign emerges in the media about the adverse health effects of smoking and the media campaign is the most striking in the attitudes of students to the treatment of the experiment. In trials, the phrase "jury sequestration" is used to prevent external factors affecting its assessment of the evidence presented. In the field of education it is not possible to "sequestrate" experiment samples as it is costly and complex to do this.

b. **Maturation.** There are students with more maturity than others in the same grades and the same age. This concept means that during the course of the experiment some children emotionally mature faster than others and this facilitates them to appreciate and better understand the material presented to them. Therefore, the maturity of the student is what influences the change of attitudes and not the experiment. This external variable is controlled by clearly establishing how the group is before the experiment. Some researchers measure aspects of students as intelligence quotients and maturity to handle this possibility.

c. **Testing.** This means the possibility of students of the experiment, after going through the experience of a pre-test, develop expertise on how to answer the post-test and this alters the actual outcome of the study. For example, in the pre-test they are familiarized with the content being measured, the format of the questions used and the expectations we have of them as a member of the experiment sample. Some researchers develop different instrument versions for pre and post tests to minimize this possibility.

d. **Instruments.** This means the possibility of errors in the instruments or the instruments used for the pre and post test doesn't measure the independent and dependent variables. Therefore there is a measurement error of the effect or the treatment. Quantitative researchers using experimental design place great emphasis on the development and validation of instruments to control for this possibility.

e. **Teacher.** This factor means the possibility that changes in students are not due to curriculum (treatment), but the teacher's ability to influence students by their personality, oration, examples used, attitudes or their sense of humor. To control this variable, educational researchers develop protocols to guide teachers and regulate their interactions with students during the course of the experiment.

f. **Relations between peers.** The process of teaching and learning is complex. It has been found that there are students who learn more by what is explained by their peers than from direct instruction of the teacher. This variable represents the possibility that the relationship between peers is the real cause of the change in an experiment and not the treatment or curriculum to which the student was exposed. As with the teacher variable, to control for this possibility, educational researchers develop protocols to guide the development of the experiment and regulate interactions among students or participants of the experiment.

g. **Mortality.** Means the possibility that students leave the group during the course of the experiment. Therefore, at the conclusion of the study the sample size ends up being very small and not representative of the population and affecting its potential generalization. Quantitative researchers are careful to select representative samples of the population studied in order to generalize their findings. Part of a representative sample lies in having an appropriate number of participants. To handle this possibility, there are quantitative researchers who select more participants in their samples in the event that if mortality occurs, does not impair the representativeness of the sample size.

h. **Differences between groups.** Means that despite selecting students at random to constitute the experimental group and the control group, the result is that the two groups are different in the comparison criteria. For example, differences in attitudes or beliefs. What some researchers do to handle this possibility is to calculate a measure of comparison for both groups, say a standard deviation for attitudes as part of the pre-test, which allows them to accurately determine the degree of difference of the groups. In the post-tests recalculate standard deviation of both groups and compare in terms of ratios, the magnitude of the changes in order to establish with certainty whether the changes observed are statistically significant or not.

i. **Rivalry between groups and changes in motivation.** To handle these possibilities, researchers, educational researchers design experiments where the only difference between the experimental group and the control group is the treatment. The most common strategies for these purposes are: 1) develop and use the same protocol to regulate the dynamics and social relations among peers and the teacher and, 2) use the same teacher with both groups.

1.4 Design Considerations for an Experimental Study

Designing an experiment involves working with the following considerations and issues: define the causal relationship to be measured, operationalize study variables, formulate hypotheses to be tested, design the treatment, sample selection, establish the validity of the study and identify ethical considerations of the experiment. Figure 1.7 illustrates this process.

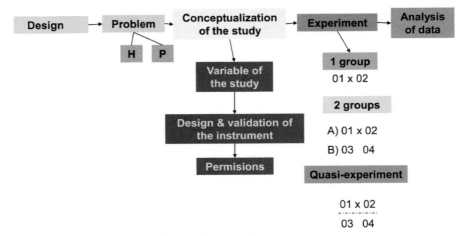

Figure 1.7   Experiment phases.

a. **Define the casual relationship to be measured.** The experimental design begins by defining what cause and effect will be tested. The relationship has to make explicit the independent variable (cause or treatment) and the dependent variable of the experiment (effect). For example, what effect has the X (independent variable) curriculum in attitudes toward smoking in fifth grade (dependent variable). The causal relationship needs to be explicit to test it with the experiment and to measure it.

b. **Justification of the casual relationship to be tested.** Every relation of cause and effect is based on a logic that explains it. These explanations are tested with the experiment in the form of hypotheses, to confirm or reject. At the beginning of experimental research in education, these explanations came from the experience and observations of the researchers. With the development of a culture of scientific research, the contemporary educational researcher is committed to use the literature to support the logic that explains the causal relationship to be tested. Literature is the repository of the work and research of published studies. There are two possibilities when justifying the relation of cause and effect that will guide the experiment in educational research. The first possibility is identified in the literature, a theory to explain the phenomenon under study and the causal relationship that will be tested. In educational research, social science theories are exported to conceptualize studies (Dressman, 2008). As an alternative if there is not a theory, the researcher must develop with literature a conceptual framework to guide their experiment and explain the causal relationship to be tested. A conceptual framework is a set of thoughts documented in the literature that relate to each other, and explain coherently how the researcher understands the research problem, the causal relationship to be studied and the hypotheses to be tested. Conceptual frameworks are constructed from existing literature. These may take the form of charts or narratives. The researcher adopts these ideas from the literature to develop their study, justify their actions and analyze their data (Ravitch and Riggan, 2012).

Theories and conceptual frameworks make the variables of the study operational and explain how the independent variable of the experiment (cause) affects the dependent variable (effect). This explanation leads to review the theory in terms of observable behaviors, factors, elements or variables measured. In the example we used, Step 1 would be to establish the explanation of how curriculum X (independent variable) changed attitudes to smoking in fifth grade students (dependent variable). The theories or conceptual frameworks are critical to developing the treatment (experimental curriculum), identify tools to measure the effect (changes in students' attitudes), estimate the time it will take to observe the change (duration of experiment), determine the manifestation of these changes (measured behaviors), and how the treatment should occur (protocol to counteract secondary or intervening variables of the study).

c. **Develop experiment treatment.** Development of treatment is to operationalize the independent variable of the study: what will be done by the treatment, how long it will last, how (procedures and protocols) will be implemented, who will implement it (personnel) and what authorizations lead to implementation. In an ideal scenario, the answers to these questions must come or be supported by the literature. Again, this is the informative function of theories and conceptual frameworks.

d. **Develop measurement instruments.** Most educational experiments rely on independent measurements of variables before and after the study. Thus one can argue whether or not there were changes in the light of the treatment. The variables or aspects that will measure the instrument must originate from the theory or conceptual framework of the study. This should facilitate identifying which instrument is most relevant and responsive to measure the rate of changes observed in the experiment.

e. **Develop experiment procedures.** The procedures are the explanation of how the study was developed. Define the sequence of events, or step by step, how, when and where the experiment begins and ends. This provides the researcher with the opportunity to identify details of implementation teams and classrooms as needed, evaluate the proposed logistics, consider alternate plans in case something does not work on the implementation of the experiment (e.g. a projector fails), define the responsibilities and expectations of the students who participate in the study, or define the responsibilities and expectations of the teachers who implemented the experiment, or estimate the implementation costs. The procedures help the researcher to identify supplemental materials of the experiment. For example, protocols for the development of the activities of the experiment, designing timetables for manage logistics of pre-tests, the start of the experiment, treatments, post-tests, data analysis and report writing of the study. In summary, the procedures provide the vision of how the experiment will be developed and this provides the opportunity for the researcher to implement their study in the reality, conditions and culture of the school or school district that is the host. Procedures become a tool for other researchers, evaluators, administrators or educators to assess and validate the study to be performed. The process is one aspect that is evaluated when validating the experiment or process authorizations with compliance committees to implement the study in the school. Finally, the procedures become discussion material when training the teachers who implement the experiment with students or to guide students who participate in the experiment.

f. **Establish the validity of the instruments and experiment procedures.** This means that the researcher ensures that their instruments and procedures meet certain criteria, standard and quality practices recognized in the scientific world. Establishing the validity of the instruments and procedures is the mechanism to ensure collection of reliable data. The validity section of this chapter addresses this issue in detail.

g. **Gain appropriate authorizations to perform the experiment.** This means soliciting authorizations with the institution where you want to perform a study. Typically, authorizations for research in educational institutions are handled through compliance offices in universities, offices of institutional research in public school systems and administrators in charge of institutions. Universities and public education count on forms and policies that outline the steps and requirements to meet.

h. Train teachers to administer the experiment treatment. This means that the researcher trains teachers to understand the purpose of the study, the nature of the experiment, the treatment used, the procedures to be used in the implementation of the study and all the details that are necessary for the success of the study.

i. Implement the experiment. Means putting the experiment into action exactly as it was planned.

1.5 Descriptive Research

The term descriptive research is used to group six designs of non experimental quantitative research which are used for the following research objectives: (a) Describe situations, current events or phenomena of education (transectional designs, surveys and longitudinal studies). (b) Determine whether situations, events or phenomena of education occur in the presence of other events, situations or phenomena (correlation), or (c) To speculate intelligently how certain situations, events or phenomena influence other situations, events or phenomena (comparison studies or ex post facto). While experimental research creates the conditions and treatment to study causal relationships and gather data, descriptive research designs study events, situations or phenomena of education in the environment or context in which they occur and manifest. The purpose of these designs is to describe the conditions, events, situations, people or trends that constitute the research problem. The other distinguishing feature of this design is that it addresses current research problems. Descriptive studies seek to describe, clarify or interpret the research problems they study. This facilitates understanding of the phenomenon and to identify and suggest (not prove) its origin, its development and its impact on other existing educational or social phenomena being studied. Descriptive studies help in three ways: make informed decisions about situations and practices that are studied, to speculate intelligently on why things are as they are at present, and identify areas where further research is needed (see Figure 1.8).

a. Descriptive research designs. There are four descriptive research designs in education. These designs differ in how to collect the data.

1. **Transectional studies.** Quantitative research assumes that the phenomena of education and members of the school community are independent entities. Therefore, it is assumed that the phenomena of education have their own characteristics that can be studied and described. This design

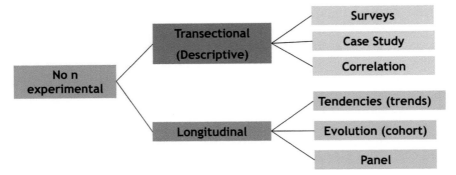

Figure 1.8   Non experimental quantitative designs.

is used to deal with these phenomena of education that are not people. In this design the research problem is studied in order to describe it in full, including its components (transectional). For example, if the study is the phenomenon of student dropouts, the researcher is required to describe the aspects that explain student dropouts. For example, what is dropout (observable manifestation of the phenomenon), what factors are triggers (reasons for it to occur), how it manifests itself (how its development is identified in the daily classes) and what are the consequences (academic, social, family). Common themes of research in transectional design are educational events (e.g., administration of standardized tests of academic achievement), processes of institutions (e.g., assemblies of fathers/mothers), or special conditions (the reality of life in the community where the school is located).

2. **Surveys.** The survey is a research design that is used to gather information from and about people: attitudes, beliefs, habits, opinions, behaviors, or feelings, to name a few. In education, the survey is used to find and describe a variety of topics from the perspective of the respondents. The hallmark of the survey is to collect data from representative samples of the population where the research question is investigated and it is used to generalize the data. Since surveys collect information from many people on a common theme for participants, the makes it easy to understand and describe the problem. Example 1 explains the descriptive nature of surveys:

> **Example 1 (Descriptive Survey).** Say a questionnaire survey with a representative sample of parents of elementary school is done. The total parents is 1200 people. Let's say that 10% of these parents are selected as a representative sample. 10% of 1200 is 120

participants. The aim of the survey is to ascertain the feelings of the sample of parents with changes in school organization, among them, extending the school day until 5:00pm. The questionnaire is administered to the sample to complete it. After tabulating the results, it is found that 70% of parents surveyed favored classes with extended hours until 5:00pm. This statement describes the feelings of the survey participants.

Surveys have the destinction that they facilitate the study of correlations. Correlation is the study of the relationship between events and conditions. In research, when an event is related to another, it is said that there is a "correlation." For example, what is the relationship between the temperature of the classroom and motivation of students to learn? Correlation means that an event tends to occur in the presence of the other. A correlation does not indicate that one event causes the other. To investigate relationships between educational events is to understand and predict these future situations where these relationships manifest (e.g. the average of high school and success in university). Example 2 illustrates the study of correlation as part of the survey design:

> **Example 2 (Correlation survey).** Let's say the researcher is interested in determining whether there is any correlation between preferences for extended hours of class and gender of the sample (father/mother). As part of their analysis, the researcher correlates the participants' gender variable (father/mother) with answers on the extended hours to determine whether there is any correlation between these two variables. Another correlation could be the age of the participants with preference for extended hours.

Surveys are commonly used in education to study the attitudes and opinions of the constituents of the school community to a theme (e.g. interest on an education policy), identify patterns of behavior (e.g. school dropouts), to assess physical facilities (e.g. feeling of constituents with physical facilities) or to identify management practices (e.g. what leadership style dominates in a school district). Surveys are common in education because of their versatility to study many different topics in a relatively short time.

3. **Correlation Study.** Correlation studies are a research design itself, which can be implemented in a leading role and not a supporting role as in surveys. The constant search for improving education, and the social nature of the profession, allows educators, administrators and researchers

to confront many dynamics leading to speculation about relationships between events and conditions. The challenge is that many of these possible relationships cannot be investigated with the experimental method due to its cost, complexity of converting to an experiment, or ethical and moral issues. The researcher has no choice but to study them through correlation studies. Correlation means that one event tends to occur in the presence of the other. Correlation does not indicate that one event causes the other. The ability to use information from correlation studies to understand the relationship between education and events, to predict future situations in which these relationships occur, makes this a very practical design in the field of education (e.g. high school GPA and success in college). Common themes in correlation studies are: (a) Human characteristics believed to be related to learning. For example, personality, self-esteem, attitudes, skills; (b) Conditions in the environment that are believed to relate to behavior. For example, standards, policies, rules, discipline, physical environment of the classroom; (c) Teaching practices and/or materials that are believed to be related to learning and, (d) Validation of measuring instruments. Although surveys are turning to the study of correlations, this constitutes a genuine design by its nature.

Correlation studies are guided by research questions or hypotheses. These are written to be answered by calculating the correlation. For example, is there a relationship between GPA and family income? If by the calculation of the correlation between variables was identifies that there is a correlation, the answer to the question is yes or a 'yes, there is a relationship between GPA and family income'. The same logic applies if a hypothesis is used to guide the study. For example, there is no relationship between GPA and family income. If by the calculation of the correlation between variables there is no correlation, the answer is 'there is no correlation between the variables'. Data in correlation studies comes from the numbers that are collected by the measuring instruments used (e.g. questionnaires, academic tests).

Example:

	Inteligence	Reading
Alicia	2	3
Pedro	1	3

Procedures of correlation studies are relatively simple: (1) **The sample/population of the study is selected.** For correlation formulas to allow discrimination, the group cannot be less than 30 people. (2) **Data collection of two variables.** Studying correlation is always two or more events understood to be related. The data is numerical scores or ratings from each of the participants in each variable. (3) **Correlation between variables is calculated: Pearsonr**. Used when you have scores of two or more variables (e.g. Scores PAEG & GPA); **Spearman** ρ (*rho*)- used when you do not have scores of variables but these can be sorted by rank (e.g. athletic ability and popularity); **Biserial (r_b)** Used when scores are obtained from only one variable and the other can be classified into one of two categories only (e.g. relationship between academic achievement (average) and family status (only child/have siblings); **Tetrachoric (*rt*)** - used when the data from the variables can only be placed in dichotomies (e.g. relationship between self-confidence (high/low) and its inclination to follow the group (big trend/little tendency).

There are two concepts which can be examined to interpret data from a correlation study: size of the relation and direction of the relation. The **size of the relation** can fluctuate from 0 (no relationship) to 1.0 (perfect relationship between the variables). The higher the correlation between two variables, the greater the reliability to make predictions of the relationship. The **direction of the relation** can be positive or negative. A +.58 Ratio is equal in size to −.58, but the relation is different. A **positive correlation** means a direct relationship, indicated by high scores on variable X are accompanied by high scores on the variable Y (e.g. High score in PAEG and GPA). A **negative correlation** means an inverse relationship, indicates that high scores in a variable are accompanied by low scores in the other variable (e.g. a lower calorie consumption and greater weight loss).

4. **Causal-Comparative Research.** The design of causal comparisons are used to study relationships of cause and effect in situations where the researcher cannot manipulate the independent variable (the cause), but the effect is obvious (dependent variable). Causation is investigated based on the dependent variable (the effect) to identify the causes (independent variable). In this sense, the comparison of causes design operates in reverse of the experimental design. In an experimental study, the researcher submits the experimental group to a treatment (the cause or independent variable) to study the effect (dependent variable). The comparison of causes design it is also identified in the literature as an *Ex post facto* research. *Ex post facto* is Latin for investigating after

the fact. For example, say students in seventh grade from secondary school X are from three different elementary schools. Teachers observe students from elementary school A excel over students from elementary schools B and C in the core subjects of English, Spanish and Mathematics. In this example, the academic achievement of students from elementary school A is visible (dependent variable or effect) to students from elementary schools B and C. The problem is that the causes (independent variable) of this outstanding academic achievement is unknown. If these teachers want to investigate this situation, they will have to study the causal relationship in the educational context where are expressed (elementary schools A, B and C) based on the dependent variable or the observed effect (academic achievement of students).

The comparison of causes design resorts to compare students, groups, schools, districts and educational systems as a research strategy. For example, in recent years there have been several studies comparing the quality of public education, primary and secondary, from several countries. To account for the quality of education and speculate on their causes, these studies collected and compared data on common criteria related to the effectiveness of teaching and the quality of education (causes or independent variables): (a) Scores on standardized tests in mathematics, science and general knowledge, (b) Teaching hours in school, (c) Study hours on assignments at home, (d) teachers' academic degree, (e) Average number of students per classroom, (f) Ratio of faculty per student, (g) Average salaries for faculty, (h) Investment in equipment and materials, and (i) Cost per student. The comparison makes it easier for researchers to identify similarities and differences. This allows speculating, from gathered data, about the strengths and weaknesses (causes) of the objects being compared.

Causal comparative studies are possible to the extent that there is data that allows for the comparison analysis. If the data does not exist, the researcher has to generate it. Causal comparative studies topics in education include: highly effective teachers versus ineffective teachers, successful schools versus schools with academic performance problems, successful students versus dropout students, communities with crime versus crime-free communities or the organizational climate of educational systems with high levels of satisfaction among teachers versus educational systems with low levels of satisfaction in the faculty.

To develop a causal comparative study, you can use research questions or hypothesis. Central to this study is to identify the causal relationship to be studied and two opposing groups to be comparable. In one group it is speculated that the causal relationship to be studied occurs and in the other that the causal relationship does not. The data can be collected from individuals, institutional statistics, available tests and any other evidence that constitutes a primary data. Two groups (classroom, school, and district) for comparison are identified. In one group the desired effect is manifested. In the second group the desired effect does not occur because it is the opposite of what you want to study. This group serves as a comparison for the study. Say the casual relationship to be studied is the academic achievement of seventh grade students from elementary school A (dependent variable) compared to students from elementary schools B and C (independent variable). Comparison criteria are identified to collect data and how these will be quantified for comparison. In our example, to determine the causal relationship, the three elementary schools A, B and C are compared on the following criteria or variables that explain the use: Curriculum content, teaching techniques from teachers, parents involvement in their children's academic affairs, the hours spent studying and ways of studying and doing assignments (individual or supervised), credentials and years of experience of teachers, the evaluations of directors on teachers. To qualify the criterion a score of 5 is assigned if the criteria meets the established standards, 3 if partially met and 0 if not satisfied.

Data from the identified criteria is collected, compared, and it is attempted to logically explain the relationship between the independent and dependent variables. Common statistics in this design are the measures of central tendency and variability to describe the similarities and differences in study groups. To accept or reject the hypothesis *Chi-squared* and t tests are used.

5. **Longitudinal studies.** Longitudinal studies are directed to collect changes that occur in individuals or in groups, through periods of three, five to 10 years as a result of the education they receive. The aim is to describe the changes and explore the relationship of these considering the time factor. The tendency is to use the same instruments throughout the study (questionnaires and interviews). Research has shown that simple changes in the questions (e.g. phrases or words) produce significant changes. Longitudinal studies are common in institutional research to determine the effectiveness of the school in achieving its mission, goals

and objectives. Three longitudinal research designs are common in institutional research:

> **Tendency studies.** These are studies researching patterns of behavior/social trends in the general population through periods of time. Population samples are studied (never the same people) when collecting data for comparison with previous studies (e.g. data on hours spent studying). The same questions are presented to the participants to study variations in the answers.
>
> **Cohort studies.** Study of the changes in a specified population (cohort) across periods. For example, what was the professional development of alumni of 1978 at the University of Phoenix.
>
> **Studies of specific individuals** (panel studies). The same individuals are studied over periods of time and try to explain the reasons for the changes.

b. **Data collection in descriptive studies.** Descriptive research can be guided by research questions or hypotheses. An example of a research question might be: What is the attraction of the University of Phoenix to encourage so many students to enroll? An example of a hypothesis would be: The main attraction of the UOP are the trimesters. Data sources of the study are identified. The primary data consists of persons directly involved with the research question and the documents available as educational policies and administrative memos. Secondary data from descriptive studies are people indirectly involved with the problem. For example, parents of students who did not experience the problem, but they heard the complaints of their children. Descriptive studies draw on a range of measuring instruments which allow quantitatively describing the problem: interview guides, questionnaires, scales, standardized tests, or any combination of instruments. The study begins by identifying and refining the research problem. Research questions or hypotheses of the study are developed. Data sources are defined (people, documents, situations). The feasibility of the research in terms of how much time it will take the survey, cost or money is needed, the usefulness of the study and the data accessibility is determined. Research instruments are developed, tested and validated. Authorizations and permissions to conduct the study are processed. Data is collected through questionnaires (regular mail, email, and internet), field observations using scales or rubrics, checklists, and structured interviews where the questions are read for people to respond. The data is analyzed and the research report is written.

c. **Data analysis.** Data analysis has three stages that are explained below:

> **Edition.** The researcher reads each data collection instrument to verify if all questions were answered clearly. There are times when the answer to a question is in another section of the instrument. In the worst case, you would have to contact the participant to answer the question. If unable to contact the participant, the researcher will have to invalidate the question.
>
> **Codification.** The researcher tabulates and quantifies the data collected by assigning a numeric code to each question. This is information that is inserted into the computer program to perform the statistical analysis.

$$
\begin{array}{ll}
\text{Single} = 1 & \text{In question 5, 24} \\
\text{Married} = 2 & \text{Answer A, 12} \\
\text{Separated} = 3 & \text{Answer B}
\end{array}
$$

> **Statistical analysis.** Descriptive statistical analysis is applied (central tendency measures or variability) or inferential (correlation, chi-squared) as appropriate to answer the research questions or hypotheses of the study.

d. **Presentation of findings.** Descriptive studies rely on three types of presentation of findings.

> **Simple data presentation.** The data is collected and organized by themes to speak for itself. For example, from the data analysis of documents or interview transcripts, documents segments or *ad verbatim* are presented in the report of findings to illustrate the issues that were found. The researcher presents the raw data so that the reader can study.
>
> **Converting Data (Descriptive Statistics).** The second strategy is to convert the data to numbers. Common statistical analyses are: measures of central tendency and measures of variability.
>
> **Thorough statistical analysis (Inferential statistics).** The data is analyzed to demonstrate how the study participants differ among themselves or between other groups. Common Statistical analysis for descriptive studies are: correlation relationships between

variables and chi-squared and T tests to establish differences between groups.

e. **Critique and challenges in descriptive research.** Most descriptive research uses people to study the research problems it addresses. This makes it susceptible to internal and external criticism. The term **internal criticism or deception** refers to the study participants providing false information intentionally. For example, the researcher is not certain whether the study participants completed the questionnaire in full awareness of the information requested by the question or lied intentionally in their responses. The researcher reduces this potential situation by maintaining a speculative and open mind to the answers and comparing their data with those provided by others to verify their consistency. Lying is due to many reasons: prejudice (political issues), personal protection or other (criminal investigations), disgust with the researcher or lack of credibility of the study or the researcher. The term **external criticism or inaccurate data** refers to the participant, despite the best of intentions; they were never clear on the facts and so providing inaccurate or misleading information.

6. **Sample Selection**

In educational research, most of the data comes from individuals or programs. In conducting research, the quantitative research model argues that it is possible to study the phenomena of education precisely, if representative samples of the population where the problem manifests are selected (i.e. people or programs). The term population of the study means possible theoretical conglomerate of all persons or programs where you could get information about the phenomenon under research. For example, if the phenomenon under study is the dropout at the secondary level in district X, theoretically speaking, the population is made up of all students who left school at the secondary level in the district. A sample means the extract of people or programs that serve as a source of information for the population from which it is extracted. If we use the example of the study of dropouts in district X, the sample would be a "reasonable" number of dropouts in the district. In quantitative research, the main concern when extracting samples of the population is to make inferences from the sample to the population studied (generalization). By selecting representative samples, the study is feasible in terms of time to perform and the fiscal cost it carries because less cases are studied, but these serve to describe the entire population. Again, the sample study is closely related to the size of the universe or study population. Identifying and selecting participants for a study is known in quantitative research as the sampling process. The sample of study is closely related to the size of the study population. The sampling process uses two parameters to evaluate its effectiveness: the sample size and representativeness. **The sample size** means that the number of participants who are selected as shown, or constitute a considerable proportion as to make inferences from the sample to the population. Table 1 shows a common table model in the literature to estimate the sample size from the size of the population where the phenomenon under study is manifested. For example, if the study population is 210 students, the sample is 136. The sample size can be estimated by the Figure 14 tables or calculated mathematically. **Representativeness** means that the sample reflects the characteristics of the population from which it is extracted. For example, an equal distribution of men and women. Let's say the population of 210 students of the example consists of 105 boys and 105 girls. A representative sample of 136 students must have 68 boys and 68 girls. The "representation" concept is relative to the characteristics of the population being studied. Representation could be understood as the nationality of the participants, their age or professional experience. It is known as a **sampling error** when the sample is not representative of the population from which it is extracted or size is not proportional to the population. Quantitative research works with two types of samples:

a. **Probability samples.** Sampling is done with the intention of making generalizations from the sample to the population studied. Probability sampling is possible when the researcher knows the size of the population, the names of its members and their characteristics. For example, the school can provide a list with the names of all students, their ages and levels of study. If this information is available, the researcher can apply any of the following selection methods. Again, the sampling probability is the primary sample and desired in quantitative studies.

   Probabilistic. The researcher assigns a number to each participant to have the same probability of being selected. A classic example of this sampling is the process of raffles. All participant names are inserted in the raffle, the tombola mixes the names and the winner is selected. Probability sampling follows

Table 1
Sample Size

Table 2.5(1)

Table for Determining Sample Size for an Evaluation
(adapted from Krejcie and Morgan, 1970)

Population	Sample	Population	Sample	Population	Sample
10	10	220	140	1,200	291
15	14	230	144	1,300	297
20	19	240	148	1,400	302
25	24	250	152	1,500	306
30	28	260	155	1,600	310
35	32	270	159	1,700	313
40	36	280	162	1,800	317
45	40	290	165	1,900	320
50	44	300	169	2,000	322
55	48	320	175	2,200	327
60	52	340	181	2,400	331
65	56	360	186	2,600	335
70	59	380	191	2,800	338
75	63	400	196	3,000	341
80	66	420	201	3,500	346
85	70	440	205	4,000	351
90	73	460	210	4,500	354
95	76	480	214	5,000	357
100	80	500	217	6,000	361
110	86	550	226	7,000	364
120	92	600	234	8,000	367
130	97	650	242	9,000	368
140	103	700	248	10,000	370
150	108	750	254	15,000	375
160	113	800	260	20,000	377
170	118	850	265	30,000	379
180	123	900	269	40,000	380
190	127	950	274	50,000	381
200	132	1,000	278	75,000	382
210	136	1,100	285	100,000	384

the same principle, the names of individuals are identified, they are assigned a number, and a lottery type selection is made to choose participants to constitute the sample by probability selection.

Systematic. Participants are selected systematically and not by chance. Participants are arranged in a list. If the population (N) is 1000 and a sample (n) of 100 then: 1000/100 = 10. Therefore, each subject in the sequence 10 is selected.

Stratified. The population is divided into subgroups: men and women. A representative sample of men and women are then selected.

b. **Non-probability samples.** It is impossible to conduct a probability sampling if the researcher does not have all the information about the study population (names, ages or levels). A non-probability sample is a sample that allows the study but not generalize the data because it does not have the information to enable sampling by probability. Non-probability samples are:

Convenience. Participants are selected for convenience, as in accessibility to people. For example, the researcher needs to survey parents about a bill that will impact education. They apply for appropriate authorizations from the educational system and present themselves to various schools to interview parents present on that day, and wish to be interviewed. This sampling does not allow the researcher to generalize their findings, but it generates information on how the proposed law is perceived.

Quota. A number of participants are selected representing a characteristic in the population under study (e.g. The syndicated teacher's feelings from a community school).

c. **Recommendations to select the sample.** Define the research problem and the population where this manifests or where it is possible to gather data. For example, if the study is on student dropouts in a school district, identify if the district or schools from that district have the names and addresses of those students. This will allow identifying the size of the study population. Gather as much information as possible from the population. This is essential to identify population characteristics. The basic information for probability sampling and non-probability is the size (number of participants) and gender distribution. List the names of people in alphabetical order and assign an identification number. Select the probability sampling method that best suits the study: probabilistic, systematic or stratified.

7. **Measurement instruments**

Quantitative research involves the use of various measurement instruments to collect data for the research problems it addresses.

a. **Questionnaires.** Are the most widely used in educational research. The questionnaire is a document (electronic or paper) consisting of questions to which the participant must answer. By their nature, can be used to collect data from many topics: knowledge (e.g. about illnesses), personal behaviors (e.g. sexual practices), attitudes and beliefs (e.g. discipline), or professional practices (e.g. management strategies). Questionnaires can be given in many forms and this makes them very versatile and practical. Questionnaires may be given by regular mail, email, internet, telephone or face to face with people. Mailed questionnaires are for people who can read and oral questionnaires are used to people who cannot read.

b. **Interviews.** In quantitative research, the interview serves as an oral questionnaire. It is commonly known as a structured interview. This oral questionnaire is developed focusing on the issues and the questions to be asked of the interviewee. The researcher reads the interview questions and the options can be selected as an answer. Sometimes interviews include one or two open questions for the interviewee to speak freely. The quantitative interview follows the same process of development and validation that is used with the questionnaire. Structured interviews are used in the field of education to collect data from people who cannot read (e.g. young children) or adult illiterates. Also structured interview is used when the researcher needs to gather data quickly and a mailed questionnaire may require more time. For example, the researcher wants to know the feelings of the residents of the community with the school and the interview turns out to be a research technique that requires less time.

c. **Measurement scales.** These are similar to questionnaires but tend to be shorter. Participants are asked to respond to questions from the instrument using a scale to rate their responses. For example, how would you describe school games day on a scale of 1 to 4, where 4 means excellent and 1 poor? Scales are used to measure attitudes, values, perceptions, behavior or professional practices. They are frequently used to study the positional values of the members of the school community or to evaluate events, situations, educational practices, group dynamics or programs. The most common scales in education are:

Summative scales. Items are developed where the participant answers in a favorable or unfavorable manner.

> Not likely 1 2 3 4 5 Likely
> I would enroll in this course

One-dimensional scale - Items are developed to measure whether the participant agrees or disagrees with the variable intended to be measured.

> In disagreement or In agreement
> Premarital sex is good practice.

Semantic differences scale. Used to measure attitudes. One or more topics to which the participant expresses his feelings commensurate with the scale are identified.

> Bad 1   2   3   4   5   Good
> Cheating on a test
> Follow directions

d. Checklists. A measurement instrument similar to the questionnaire, the only thing is that all statements are answered with a yes or no, has or has not, met or not met. Used to measure or evaluate situations or specific aspects of education as would be management practices. The scales are used in two ways. First, like an instrument the participant fills. For example,

> Your school has:
> Sport facilities   yes___ no___
> Parents Committee   yes___ no___

The second alternative is when the instrument is completed by the researcher. For example,
The school complies with the following requirements:

> Clean facilities   yes___ no___
> Facilities in good condition   yes___ no___
> Security Guards   yes___ no___

e. **Evidence of student development.** Educational researchers collect data through standardized tests that help to understand the changes that occur in students as a result of the education they receive. This instrument is also used to understand the achievement of educational goals and objectives.

> **Achievement tests.** Tests that measure the cognitive development or gain of students. These tests differ greatly. There are tests to measure general knowledge or expertise in specific areas. They are used to measure the cognitive development of students. Since the 1980s, they have been used to determine the effectiveness of educational institutions and their faculty in promoting student learning.
>
> **Psychological testing.** Standardized tests to measure aspects such as personality, self-esteem or the level of intelligence of the student. There are projective tests where the participant interprets a portrait, a drawing or relationship between words in sentences. The emotional development of the student is considered to be a fundamental aspect of education. Such tests are used to place students in academic levels in primary or high schools or as a benchmark to determine the emotional development of the student as a result of their education.
>
> **Physical attributes tests.** Tests to measure motor skills, physical fitness, strength, speed or reaction time. They are frequently used from preschool through high school classes in physical education, adapted physical education and special education. They are used as tools for diagnosis, development and evaluation of progress at the time of measuring the achievement of educational goals and objectives.

8. **Validity and reliability of data**

All research to be considered scientific must show that the information collected is reliable and valid. **When researchers talk about the validity of data it is saying that these closely correspond to the situation investigated.** The validity of quantitative studies focus on the instruments used to collect the data, the procedures used to conduct the research and the form used in the analysis of the data collected.

a. **Instrumentation validity.** Refers to whether the instruments used in the study measured or collected the data they were supposed to measure or collect. Three techniques can be used to establish the validity of the measurement instruments.

Content validity. Refers to the instrument appearing to measure what the researcher intends to measure. In many questionnaire studies, where the researcher designs their instrument, this is the only type of validity that is discussed. The researcher constructs the questionnaire and submits it to a group of experts to indicate whether the instrument measures what it appears to measure. This technique of validity has the advantage of being a quick validation. The technique is considered less credible because it is based on the opinion of people.

Criteria validity. Refers to whether the data collected with the instrument are consistent with data from other similar instruments for the same population. For example, if the results of the mathematics test developed by School X are similar to data from the standardized test of the Department of Education, then it could be argued that the school X instrument measures what it is supposed to measure because it produces data similar to an instrument already established. This technique creates a great sense of trust to the researcher because the instrument generates similar data to other known instruments. In many theses and dissertations this technique is not always possible to use because there are no instruments on the topic of study.

Field validity. Is known as a pilot study and allows testing the instrument in the field to determine whether it generates the data needed. The researcher administers the questionnaire to a similar study population sample and asks participants to react to the instrument (e.g. if they found the questions difficult, understood each question or appearance). The researcher tries to determine with this exercise if the message that he/she wishes to communicate in each question is the same interpreted by the participants who read it. This exercise also serves to test the proposed statistical analysis for the study and see if the required data is generated to answer the research questions or test the hypotheses. It is a technique that helps the researcher to understand and refine the instrument before performing the study. It may involve additional work for the researcher to identify participants for the pilot study, but the ability to understand and improve the instrument before the study deserves any effort.

Instrument reliability. Reliability refers to whether the instrument is consistent in measuring what it is supposed to measure.

For example, say you are trying to determine your body weight on a scale in three separate occasions and get different measurements. Probably, the reaction will be to not rely on the scale as it is not consistent in measuring body weight. The scale is unreliable. This same logic applies to the research instruments. Three techniques are used to establish the reliability of the research instrument.

> **Pre-post test.** The instrument is administered at two different times to the same population to estimate the correlation between the first and second administration. If the instrument measures what it is supposed to measure, and participants respond with good conscience, it should produce similar results in the two administrations.
>
> **Split-Half Method.** An instrument with two groups or sets of questions that measure the same thing is constructed (even and odd questions). For example, the even questions ask the same information as odd questions, but using different words. The instrument is administered and the scores of the odd and even items are determined. Correlations are estimated with these scores. The higher the correlation, the more reliable the instrument is considered because the even and odd questions reflect similar scores. The rationale behind this technique is simple, if odd and even questions seek the same information, and if the participant answers with good conscience, both sets of questions should produce similar responses.
>
> **Equivalent forms.** Two equivalent versions of the instrument are given to the same population. For example: two versions of the same test are developed. Results are compared to determine whether they are similar or divergent. If the instrument measures what it is supposed to measure, and participants respond with good conscience in both versions, similar data must be generated. This similarity is considered evidence of the reliability of the instrument.

b. **Procedure validity.** Refers to other procedures used in the investigation being in line with the study objective. This means that the way of conducting research does not alter the data of the phenomenon being studied. In other words, conducting the study does not affect the way the participants act or think. Quantitative research establishes the validity of the process through the development and validation of protocols. A protocol is a guide where the researcher outlines the steps of the study in its implementation, instructions or commands told to the participants and make explicit the role as a researcher. This protocol is submitted for consideration by experts to react, to assess and validate it. Once this process is complete, the researcher conducts their study and is guided by this protocol from beginning to end.

    c. **Data analysis validity.** Refers to whether the techniques used in data analysis are sensitive to the information gathered in the study. This means that the techniques do not produce a false impression of the data.

9. **Data analysis**

In quantitative research, data analysis lies in the use of statistics as a tool for organization and interpretation. Statistics is the science that deals with the organization, analysis, and presentation of numerical data as a basis for the description and comparison of facts observed through research developed. The statistics are grouped into two categories: descriptive and inferential. Statistics allow the researcher to summarize, describe, evaluate, interpret and communicate numerical information from individuals or groups. In this technological age, there are computer programs that perform the calculation of the analysis. The main challenge for educational researchers is no longer calculating the mathematical result on paper and pencil, but to identify the statistical analysis that best meets the objective of the study and interpret the results generated by computer programs. Some computer programs interpret the mathematical findings leaving the researcher to reflect of the implications of these findings. A detailed discussion of statistics in quantitative research is beyond the scope of this chapter. Statistics is a discipline of study and extensive in its own merit not to be limited to a few paragraphs.

    a. **Descriptive statistics.** The aim of descriptive research is to describe the research problem under study. Two types of descriptions are common: describe the study participants as individuals or as a group. Common statistical analyses in descriptive research come from descriptive statistics. Descriptive statistics consists of statistical analyses that help to summarize data and describe the characteristics of the problem or phenomenon under study.

The three most common analyses of descriptive statistics are counting frequencies, measures of central tendency and measures of variability.

b. **Inferential statistics.** Inferential statistics are statistical analyses that are used to make decisions on the acceptance or rejection of hypotheses and to make inferences about possible relationships or differences between two or more variables. Statistical inference is organized into two main categories:

> **Parametric statistics.** Statistical tests that are used to establish the significance value of the hypotheses under conditions where the researcher knows that the study sample was probabilistic and survey data are of interval type.
>
> **Non-parametric statistics.** Statistical tests that are used to establish the significance value of the hypotheses under conditions where the study sample is not probabilistic and the study data are nominal or ordinal.

Much of the effort of quantitative research is in determining causal relationships. For these purposes it uses experimental, quasi-experimental and causes comparison designs. The aim is to compare groups to understand the relationships between the variables under study. By studying two groups or more to interpret the behavior of the variables, the following statistical tests are used to interpret the data and show changes between experimental and control groups: Chi-squared, *t* tests and variance analysis. To determine whether the causal relations are really relations of cause and effect, researchers turn to the hypothesis as a working tool to confirm or reject them. A **hypothesis** is a tentative answer to a causal relationship of the study. For example, children who participate in sports develop more strength than children who do not participate. In this example, the hypothesis states that participation in sports helps children develop strength. Researchers rely on two types of hypotheses when studying causal relationships:

> **Alternative hypothesis.** They are statements about the world or phenomenon under study. The alternative hypothesis can be of two types: directional or bidirectional.
>
> > **Directional (*one-tail*).** Hypothesis that establishes the relationship between two variables and specifies the direction in which the relationship occurs. For example,

children who participate in sports develop more strength than those not participating. The hypothesis clearly establishes the causal relationship.

Non-directional (*two-tail*). Hypothesis that establishes a causal relationship, but does not indicate the direction. For example, there is a significant relationship between television viewing and academic achievement. This hypothesis states that the hours spent watching television influences the academic achievement of students. The hypothesis does not indicate how the causal relationship is. For example, if more hours of television shows better academic performance or vice versa. The non-directional hypothesis is used when the researcher suspects that there is a causal relationship, but does not know how this relationship or its direction is.

Null hypothesis. This hypothesis establishes that there is no relationship. In hypothesis-driven research, the null hypothesis is important because it is the hypothesis consistent with the term "probability" or chance. **Inferential** statistical analyzes orientation is to estimate probability. Statistically, it is the hypothesis that the researcher seeks to reject. By deduction, to reject/accept the null hypothesis implies to confirm /not confirm the alternate hypothesis.

The null hypothesis means that the researcher will seek the most efficient way to reject the hypothesis of the study. The most efficient way is to prove the opposite (null hypothesis). For example, say the study's hypothesis is that swans are white. The researcher can collect a million white swans but this does not prove that they are all white. So proceeds to test a null hypothesis... identify a swan that is not white. As soon a swan that is not white is identified, the null hypothesis is confirmed and the hypothesis of the study is rejected. Figure 1.9 presents a normal distribution curve showing both extremes (shaded area).

When the researcher uses a directional hypothesis, what is establishing is the statistical analysis to use, look for the swan that is not white on the right end of the distribution curve. When using a non-directional hypothesis, this indicates looking for that swan that is not white at both ends of the

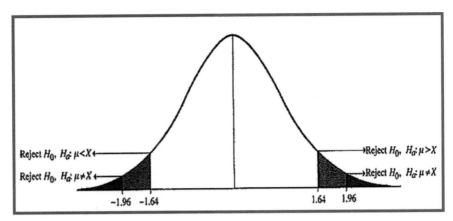

Figure 1.9   Distribution curve and hypotheses.

distribution curve. To use the null hypothesis, the researcher has to set the parameters that they will be using in terms of statistical probabilities of the data for the appearance of the swan that is not white and then be able to reject or not the null hypothesis.

Example. Say you design a study to determine whether there is a relationship between hours spent watching television and academic achievement (non-directional hypothesis). The researcher defines its parameters as follows:

Figure 1.10 shows that, the curve parameters are set in terms watching TV more than two hours and less than two hours of television daily to determine whether there is any relationship to academic achievement. Data is collected and seeks to identify whether there is correlation between the number of hours devoted to television and academic achievement.

In interpreting the data to accept or reject their hypotheses, the researcher faces the possibility of committing two errors:

Type I error. Reject the null hypothesis (accept alternative hypothesis) and conclude that there is a relationship between two variables when in fact there is no relationship.

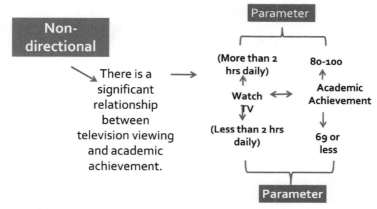

Figure 1.10 Hypotheses.

Type II error. Accept the null hypothesis (reject the alternative hypothesis) and conclude that there is no relationship between the variables when in fact there is a relationship.

The most common reason for Type I and II errors is meddling with **secondary variables** in the study design. This occurs because the researcher fails to control all possible variables of the study. The researcher believes that changes in the dependent variable (what you study) are due to the independent variable (what causes the change) when it is not. This error is corrected by paying attention to the study design.

From Hypotheses to Statistically Significant. The term "statistically significant" in research, refers to the probability of something happening and not that the finding is important. A "statistically significant" finding means that the relationship between two variables occurs 95% of the time. In other words, the researcher has a confidence level of 5% of making a Type I error (accepting null and rejecting the alternative). A good study indicates the statistical criterion used to accept or reject the hypothesis. In other words, what is going to be considered "a statistically significant finding"? The most common significance level in educational research is .05. This means a chance of making a type I error is 5 in 100. Other levels are:

1. .01 = probability of 1 in 100
2. .001 = probability of 1 in 1000
3. .10 = probability of 10 in 100

"Power" Concept in Statistics. There are hundreds of statistical analyses. How assertive is the researcher to select the best statistical analysis for your goals, determine the accuracy to establish relationships or differences between variables. The term "statistical power" refers to the properties of a statistical analysis to become more appropriate than another given the circumstances. A statistical analysis is considered more powerful than the other when considering all values or involved cases (e.g. standard deviation vs. range).

10. **Closing remarks on statistical inference.** There are times when researchers want to know how to compare the data from their study with data from other studies. The way to determine this is to place the data on a graph to determine its configuration. Usually, data for analysis of a variable, that are similar, tend to group into clusters and dissonant cases away from the group. In theoretical terms, people have characteristics that they have in common. The elements that deviate from the "norm" represent the dissonant cases. The distribution curve identifies whether the data from the study, translated into a graph, forms a normal curve or not. Below is an example of a normal distribution curve and potential analysis that the researcher can perform with their descriptive study data. The center of the distribution curve is identified with zero (0). Then there is +1, +2 to the right and −1 or −2 left. The center means that common characteristics of a variable are grouped or are positioned in the center of the distribution curve. These characteristics are referred to as the norm. The center is considered the norm for the human characteristics. For example, say that the variable is the intelligence of the participants. Most people have an IQ between 100–110. So the people who have an IQ of 110–120 mathematically fall into the category of +1 standard deviation to the right. By being more intelligent, deviate from the norm. The same applies if they have an IQ of less than 100. Researchers can estimate and compare the study participant with an estimated standard (normal distribution curve) through their study scores (Z scores) or by comparing the scores of the participants as a group (t scores). (See Figure 1.11).

11. **Presentation of quantitative data.** Quantitative studies rely on three strategies to present their findings: using tables, graphs and narrative.

   a. **Tables.** The table is a strategy to provide summary quantitative data without comparing. A common table uses horizontal or vertical columns, as required by the researcher, to display quantitative data.

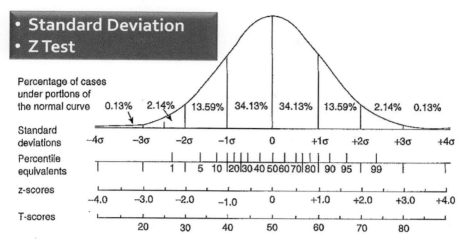

Figure 1.11   Normal distribution curve.

Some researchers use a table to display the data by research question or hypothesis. Others use tables to summarize the data for the study variables. By its nature, the table provides summarizing and presenting data without comparison.

b. **Charts.** The chart is a strategy to present/display quantitative data summaries, but requires a comparison of these by nature. The most common charts are pie charts and bar graphs. The nature of circles or bars to display data allows summarizing and imposes a visual comparison of these.

c. **Narrative.** The narrative is a complementary technique to the tables and charts. As a complementary technique consists of written comments or statements to review the most significant data of tables or graphs.

2

Qualitative Research

Omar A. Ponce and Nellie Pagan

Metropolitan University, PR, United States

2.1 Description of Qualitative Research

Qualitative research is the second model of scientific research used to study the problems of education. Arriving in the field of Social Sciences and Humanities Education (Pauls, 2005). The term "**qualitative research in education**" is used in the literature to describe a range of field rearch that exhibit two distinctions (Ponce, 2014): the study of social phenomena and problems of education and the approach to these from the perspective of those who live and experience them. We will examine these distinctions to understand qualitative research as a model of educational research:

a. **Emphasis on research of social phenomena of education.** Education is social. It occurs in educational institutions and develops interactions between people and objects. Human interactions refer to relations that occur between teachers, students and administrators. Interactions with objects refer to relationships that occur with the rules, policies, goals and educational objectives and curricula. From these interactions, phenomena and problems that affect the effectiveness of education are generated. **Social phenomena and problems** of education means, those behaviors, situations and dynamics with more than one root cause, more than one solution and more than one way to be interpreted and explained. For example, Why students in the same classroom find their teacher fascinating, while others consider them to be boring? Why students learn better with some styles of teaching and others not? Why some students persevere in their studies while others, under the same conditions of life, leave the school? Why there are schools that achieve extraordinary accomplishments while others do not, despite following the same curriculum, have teachers with the same credentials and a similar student profile? Qualitative research in education studies social phenomena of the profession because it argues that human behavior is the result of the interpretation that people make of their real life. The premise is that people interpret things that happen because they ascribe to these values and emotions. If these interpretations (causes) are considered, then one can understand the behavior of individuals/groups and intervene with them (effects). If this view of human behavior is accepted, then it is easy to recognize and understand the complexity of the multiple relationships and social interactions in education systems by interpretations that exist of how it can be or should be the field of education. From this perspective, one can understand that there will always be a more effective way to teach, learn or administer an educational institution. This is important to understand that because of these multiple interpretations many of the "problems and social phenomena" of education arise. For example, a student may be fascinated by the way a teacher teaches physical education and another not. The interpretation of the student in their physical education class can affect their interest in this field of study. That interest can influence their learning and even lifestyle that they eventually develop. What this means, if I am interested to improve my effectiveness as a teacher of physical education? What does this fact mean, if as a director I have to assess the accomplishments of a teacher and the scope of student learning? From the foundation of qualitative research, this means that to understand and intervene with the operation of a school, a classroom or a student, it is essential to understand how their constituents (students, teachers and parents) interpret and act on events and situations that occur there. An educational event can have more than one interpretation and cause many actions that affect student learning and the effectiveness of the educational institution. For example, the school principal may provide a guideline that is welcome by teachers, but causes upset to parents and students. As simple or complex as it may seem, the phenomenology of the experiences and lessons is a way to enter these social phenomena of human behavior that occur in educational settings.

b. **Research from the perspective of those involved in education.** Qualitative research in education seeks to "**understand**" **social phenomena and problems of education through the experiences and interpretations of those who live and experience them**. In other words, **the phenomenology** of the experiences and interpretations of people become "windows" to enter those social phenomena that exist in the field of education. Phenomenology is a philosophy that argues that humans know our social world through the senses and experiences. When the phenomenology of human behavior leads the educational research, options and opportunities to research and understand the range of social phenomena in this field are created. Qualitative research of social phenomena of education, contributes to the understanding of many human interactions and relationships that occur in educational settings. This is critical when developing intervention strategies for teaching, educational materials or educational policies and advance scientific knowledge about this profession.

This results in two forms of inquiry. The first form of research is direct observation of human behavior, its dynamics or phenomena as they occur in educational settings. This allows the researcher to determine, for themselves, the context or educational event that occurs, the resulting interpretation and how they affect one another. The second form of research is when the researcher is unable to witness the phenomena studied because it has already occurred. Then, there is no choice but to study and understand human behavior, its dynamics and phenomena through the experiences spoken of those who live and experience it. For example, interviewing defectors to try to understand why students left school. The aim of qualitative research in education is not to measure the phenomena it studies, like physical entities, but study them to understand and describe them as social beings. With that understanding, we can make tangible the social phenomena studied and discuss them with others scientifically. This implies that qualitative research must be field research for the researcher to observe first hand, to interview or analyze, how the environment, the interpretation of the participant and their behavior are understood in one context. Qualitative research is inductive and the explanation of the phenomenon under study is the expected product. As qualitative research seeks to understand the phenomena studied, the data tends to be descriptions of observations, verbatim interviews or information from the analysis of documents, objects or pictures (qualitative data). The qualitative research report tends to be flexible in its format and in its less technical scientific vocabulary, because it is an explanation of the phenomenon being studied and in its natural context.

2.2 Models of Qualitative Research

Three models of qualitative research are visible in the field of education: the constructivist/interpretive model, the transformational/social critique model and the realistic model.

a. **Constructivist/interpretive research.** This type of qualitative research is based on phenomenology and symbolic interaction philosophies. Phenomenology and symbolic interaction brought to educational research emphasize the importance of understanding the interpretations of the constituents and their social interactions to enter and meet the educational work. The focus of constructivist or interpretive research is the understanding of "social phenomena" of education or the study of social interactions that occur in educational settings by diverse values and cultures of their constituents. In this category there are phenomenological, ethnographic, grounded theory and case studies. These qualitative studies were the first to be used in the field of education. They are known as constructivist or interpretive studies, because the understanding of the phenomena under research is the construction made by the researcher of the interpretations and experiences of the study participants.

b. **Transformational/Social critique research.** Transformational studies recognize that the interpretation by people determines their reality (phenomenology). It is argued that this interpretation is affected by other factors that must be considered when studying educational settings (Symbolic Interaction). These factors include the gender of the person (theories of gender), political influence governing educational institutions (theory of social critique), and new visions of life and social orders that have been developed in contemporary societies (postmodernist theory). The focus of transformational studies is to expose the voices of those which are recipients of administrative decisions, policies and hierarchies of educational systems. This category includes studies of social critique (study on the political impact on the constituents and their problems), feminists (the educational reality from the perspective of women), postmodernists (new worldviews and social orders in educational systems) and action research (the involvement of participants in a study on the solution of the research problem). They are known as transformational studies because the goal is to use research to expose the voices of those who are administratively "marginalized" or "disadvantaged" by the political interests of those who run education systems. The purpose of these studies is to produce changes in educational systems by exposing the consequences of educational policies and administrative decisions. This type of study relies solely on the experiences of those marginalized or disadvantaged groups in education systems. These studies have generated much controversy in the field of education and have been excluded from participating in federally funded projects for its political content and are not considered scientific (NRC, 2002).

c. **Realistic research approach.** This type of qualitative research is based on philosophical principles of realism. Realism, in its many philosophical variations acknowledges that social phenomena exist in the minds of people (phenomenology and symbolic interaction) and in the external reality to the human being (realism). It postulates that there is an external social reality to the interpretation of human beings. The interpretation is not only a reflection of the human mind. It is an evaluation that people make about external phenomena. Social reality lies not only in the minds of people as if it were a world of ideas. Social reality is the result of interacting with external reality. Is captured through the senses and language, when able to perceive and identify. External reality affects the individual, but the human beings also affect that reality with their interpretations and actions (Pring, 2000). In the field of education, qualitative research approaches are identified under realistic philosophies of pragmatism (Patton, 1980), post-positivism (Creswell, 2007) and transcendental realism (Miles & Huberman, 1994). The focus of qualitative studies with approaches to realism is to capture the external social reality through the experiences of those who live and experience them. In this way it seeks to identify patterns of expression in the phenomena being studied and to generate verifiable knowledge (Miles & Huberman, 1994). In this type of qualitative research, the researcher transcends the experiences of those who experienced the phenomenon, to find the objective aspects of the social phenomena studied. This type of qualitative research is distinguished by adhering to systematic and rigorous procedures as in quantitative research. Creswell (2007) indicates that this type of qualitative research is seen in researchers whose alignment is in quantitative research.

2.3 The Mentality of the Qualitative Researcher

The act of research is grounded in the scientific thinking of the educational researcher. Understanding this mindset is critical in qualitative research because the researcher is the research instrument. It is the researcher who enters the field to observe or interview to understand the phenomena studied. Qualitative research is arduous because it is field investigation. It is a fascinating model for those who have the "intellectual curiosity" to inquire into human and social complexity of education. It is an appropriate model for those educational researchers who have the "personality" and the "tact" to interact with people. Qualitative research is not the artificial application of a research method to a problem. Qualitative research is an "art and a science" that develops with the practice of observing and interviewing. Hence the name of naturalistic research that some authors ascribe (Lincoln & Guba, 1985). We'll examine the personal and professional qualities that define the mentality of the qualitative researcher:

a. **Qualitative thinking.** The mentality of a qualitative researcher is qualitative thinking (http://www.csse.monash.edu.au/smarkham/resources/qual.htm). Qualitative thinking consists of the mental ability to identify the qualities, characteristics, or distinctive characteristics of these social phenomena under research. These qualities stem from the behaviors, values, experiences and emotions of the study participants. There is no magic formula to identify these. Many think that qualitative research is limited to qualitative research interviewing/observing and reporting the data. The fact is that the qualities of a social phenomenon emerge from the researcher's ability to identify these through the experiences of those who live them. These are not always reported as they may not be aware of them. For example, if we study the effectiveness of the game as a teaching strategy in preschool children, the probability of effectiveness comes from understanding this behavior and emotions observed in the children. It is likely that children cannot verbally express the phenomenon they are experiencing with their limited vocabulary and mental maturation. There are occasions where people experience a phenomenon, but cannot define or explain it categorically. It is for this reason that the qualities and properties of phenomena emerge from the researcher's ability to identify and describe them.

b. **Intellectual curiosity and scientific thinking.** The act of studying social phenomena is always guided by the intellectual curiosity of the researcher or questioning the why of things. This is the role and importance of the research questions of the study. The act of qualitative research is not always limited to the experiences reported by those who live them. Studying education qualitatively involves digging into these dynamics, that not even the same people that experience them understand sometimes. Investigating qualitatively involves constant reflection to give order to information that is collected about the phenomena being studied. That is, to understand the experiences of those who live them, how they originate and what are the consequences.

c. **Tolerance for ambiguity**. When social phenomena are studied, patience and tolerance for ambiguity is needed. The problems and social phenomena of education can be characterized by its lack of clarity and states of confusion and uncertainty among those who experience them. Sometimes the emotion that circumscribes an event makes the work of the researcher more difficult. Hence the need to investigate them qualitatively. The patience of the researcher to enter this uncertainty, understand and organize mentally presents a large exercise in patience in the process of observing or interviewing. It also entails tolerance of ambiguity in the search of data to understand the phenomenon. The educational researcher who cannot handle ambiguity, while making roads where there aren't any, will have many difficulties with this research model.

d. **Respect for cultural diversity.** Qualitative research seeks to understand social phenomena of education through the experiences of those who live them. These life experiences are part of the values, beliefs and culture of the people. To understand these, the qualitative researcher needs to venture into the cultural diversity of the people that have become participants in their study. Ethically speaking, and to research responsibly, the qualitative researcher has no right to morally judge the values and beliefs of the people in their study. Their responsibility is to understand the phenomena studied in the context of the values of the people, to communicate scientifically, ethically and responsibly in solving the problems of education. The qualitative researcher has to be a person of tact, good manners and responsive to the educational realities they investigate. Interacting with people implies tact, good manners and respect for human dignity.

e. **High ethical, moral and legal sense of the research.** All educational research must adhere to the legal requirements established by education systems where the study will be conducted. Generally, these requirements focus on obtaining authorization to conduct the study in the school/school district and the written consents of the teachers, students or their parents, being the people who constitute the study participants. Another important aspect is the confidentiality of the information gathered through observations or interviews. Responsible management of information has a high ethical, moral and legal sense of research as a scientific and professional practice. Another fundamental aspect is the researcher's responsibility to report accurate, precise and exact positions, experiences, values and beliefs of the participants in their study and the phenomenon studied.

2.4 Generating New Knowledge Through Qualitative Research

Educational research is understood as the scientific search for evidence-data to generate knowledge to help solve the problems of education. The word "knowledge" means many things: comprehension, understanding, discernment, judgment, information and erudition. In qualitative research, knowledge is a product of the mind because it is a building or a mental exercise. Knowledge is "knowing the essence" of the phenomena investigated. Knowledge is generated in a relationship between the researcher and the phenomenon under research. The phenomenon under research can have tangible and real properties (e.g., the temperature of the classroom) perceived or intangible (e.g., if the room temperature is pleasant or not). By the nature of the problems/social phenomena that qualitative research addresses, the knowledge that a researcher generates is the formation of an idea or image of the object studied (understanding). This knowledge is real when it corresponds to the object represented or under research. **Describing and explaining a social phenomenon, as do qualitative researchers, is not the interpretation of the researcher, but the result of their research. The authenticity of this knowledge lies in the correspondence between the mental image formed by the reseracher and the investigated object or phenomenon.**

For qualitative researchers, the best way to achieve this mental picture of the problem, and **validate** it, is entering into educational settings (field research) where the problems investigated occur (context). Once in the field, these are studied to understand them clearly to describe them in detail and explain them. **It means the act of qualitative research is not only to understand the phenomenon, but also includes the exercise to establish and verify the correspondence between the mental image constructed and the phenomenon under study in order to argue about the authenticity of the findings and contribution made to the field of knowledge.** It also means to observe with their own eyes, in the front row, the phenomena studied, or in face to face interviews with people living the phenomena researched, representing the best research techniques to understand social phenomena and human interpretation occurring in the field of education. The great challenge that involves validating the data of the study is to identify the observable manifestations of the social phenomenon under research. For example, Jean Piaget, the French psychologist, developed a theory of how learning in children occurs. The explanation of Piaget argues that children go through a number of mental stages from acquiring information to using it with dominance.

Mental stages of the theory cannot be observed directly in the brain (physical reality), but they can be linked to observable behaviors that facilitate assessing children's learning in the light of his theory (social reality).

2.5 Qualitative Research Designs

To study social phenomena of education, qualitative researchers use a variety of research designs. Five designs dominate qualitative research identified as relevant in the field of education (Ponce, 2014: Licthman, 2011; Mertens, 2005; Freebody, 2003; Merriam, 1988; Woods, 1996; Bogdan & Biklen, 1992). Understood by relevant, responsive and allow the study of the situations and characteristics of the profession. The qualitative research designs differ from the viewpoint that they adopt to study various social phenomena of education. For example, educational researchers use the experiences of the constituents of the schools to understand the phenomena investigated (phenomenology), others prefer to study the culture and context of the phenomenon (ethnography), and others address aspects of educational policy to study the research problem (social critique). The viewpoint of the design defines the focus of the study. This helps to outline the next steps in addressing the social phenomenon under study. We point this out because using qualitative research designs are not necessarily translated into a sequence of steps that are learned and applied mechanically. Understanding this helps to use the design and viewpoint effectively. This helps the design reflect the particularities that facilitates defining it as one design or another. The confusion of designs, or ambivalence of features in the implementation of research, it is noted in the literature as the origin of much methodological confusion (Ander-Egg, 2003: Yin, 1994).

   a. **Phenomenological research.** Phenomenological research studies social phenomena of education through the "experiences" of those who "live and experience it." The experiences become the "window" to study, explore, understand and describe the phenomenon under research. According to Van Manen (1990), phenomenological research consists of the following activities: the study of human experience, as experienced, the explanation of the phenomenon under study, as manifested in the human consciousness, entering into the essence of the experience and therefore the phenomenon under investigation, and description of the meaning of the experience for the person. Phenomenological research seeks to first understand the experiences of each person who constitutes the study sample. It is common in phenomenological researchers to create individual profiles of the phenomenon, as experienced by each study participant. As a second step, seek to understand and identify the common elements of the phenomenon that is manifested in all study participants. For example, what is the essence of the phenomenon? What are its components? How does it manifest? According to Merriam and Associates (2002), by doing this, the phenomenological researcher assumes that there are common elements of the phenomenon and that all those who lived also experienced it at some point. These common elements are the essence of the phenomenon. Thus, the aim of a phenomenological study is to identify and describe the essential aspects of the phenomenon under study: its structure, its components and its manifestations.

The phenomenological design is relevant in educational research by the amount of individual and social phenomena manifested and where mediate the interpretation of their constituents. Examples of individual phenomena are motivation for learning, defection, retention and student persistence, test taking anxiety and learning itself, to name a few. When these phenomenons are evident in the collective, they influence the effectiveness of the educational institution. If these phenomenons are not studied and understood, then no one can intervene with individuals or with groups, where they appear. The product may be a phenomenological study profile, an explanation or a model of the phenomenon being studied. That profile, explanation, or model of the phenomenon has to allow understanding of how it is and how it manifests itself in the educational scenario. The development of profiles, explanations or models of social phenomena is essential when developing educational strategies to intervene with them to develop educational policies that address them or develop training for the constituents of the educational system to understand and handle them. Profiles, explanations and models allow in turn generation of other studies to provide continuity and generate new understanding about the phenomenon.

b. **Ethnographic Research.** Ethnography is the study of the culture of a group: how it is, what unites them, how they relate and how they work. The goal of ethnographic research is to describe in detail the cultural aspect that it researches: beliefs, values, customs or behaviors. Because "culture" is a multidimensional and complex phenomenon to study, many ethnographic studies in education are limited to a single aspect, as would be a belief. Ethnographic research comes to the educational field of anthropology. Anthropologists have a long history of going to live with unknown tribes and become one of the group to understand the culture. Ethnography involves an extensive immersion in the culture being studied to gain knowledge through observation, interviews with native culture and the study of documents or artifacts generated by the group. Gaining knowledge of the culture means to produce a cultural "portrait" from the perspective of the "native" group (emic perspective). Ethnographic "portraits" join together and integrate the viewpoint of the researcher as a research instrument (epic perspective). Education systems are composed of people of various levels, beliefs and customs. When all these people coincide in the same educational setting and work toward achieving educational goals, guided by institutional rules they generated dynamics that lead to subgroups within the school and that shows the "culture" that forms in the environment. These interactions are colored by the values, beliefs and cultures that each of its constituents bring to the educational setting. Ethnography has been used in education to understand the cultures of classrooms, schools, sports groups, communities of origin of the students, or the organizational climate of schools or educational systems. This type of study is important in order to learn the cultural elements of an educational system or group. This understanding is essential to develop sensitivity to cultural diversity in the education system. This sensitivity may take the form of academic policies or educational approaches to cultural diversity as would be the outline teaching strategies. It could take the form of aligning curriculum to the cultures of the constituent contents; assess the extent of school policies and regulations and to develop regulations that are sensitive to the cultural diversity of the constituents. The product of an ethnographic study can be theories or models illustrating these valuation and cultural relations of cause (interpretations) and effects (behaviors). They can also be cultural profiles or radiographic values of those individuals or groups that exist in the educational systems.

c. **Case Study.** This design consists of a deep study of a "case". The case can be a child, a classroom, a school, a community or an institutional process. The "case" is studied because it exhibits a feature that requires researching to improve education. As a phenomenon of study, the "case" is complex because of the many factors that clearly manifest within it. Therefore, the aim of the research is to study the "case" in all its dimensions and extent to understand it. The case study uses observations of the case, interviews and document analysis. In this design, the researcher can venture into the real scenario where the case occurs and appears to be an ethnographic study. They can examine documents and interview and may appear to be an historical or phenomenological study. The virtue of this design is that the researcher has the flexibility and amplitude to inquire into "all aspects" of the case that are necessary to be understood. For example, if it is phenomenology, the study focuses on the experience of the participant. If the study is ethnography, the focus of the study is the "culture". When a researcher argues that he\she is doing a case study, not only is commited, but it is expected of him or her to approach to the "case" from multiple perspectives and depth. The ability to approach the case from multiple perspectives is the hallmark of the case study as a qualitative research methodology. The case study has been widely used in education to understand different phenomena. For example, the study of children living in poverty and cultures who persevere in school despite family/social adversity and manage to leave that orginal environment. The study of educators who achieve impressive academic development in their students. The study of second language acquisition in foreign children or highly successful schools. The case study also has been used extensively to study institutional or educational processes. For example, the processes of acculturation experienced by foreign children joining schools in other cultures, the transitions processes between educational levels or when students undergo or are subjected to a curriculum or academic program. The phenomena and processes of this nature become "cases" to study everything they offer to improve education. Case studies are used to understand phenomena, reconstruct transitions and educational processes that occur, which are not seen in the physical world, but by students who experience and live them. This allows us to identify how they occur and how are these phenomena. This helps create the conditions of physical infrastructure, student counseling and guidance on the skills that are needed to work to get through them successfully.

  d. **Grounded Theory.** The objective of this design is to generate a theory that explains the phenomenon under study or some aspect of the data generated from the research. For example, a theory that explains the phenomenon of student persistence. The study begins by collecting data on the phenomenon under study. The data is studied, explanations and manifestations of the phenomenon are generated, and more data is collected to confirm these explanations, until it is possible to develop a theory supported with data. The study concludes when the "information overload" point is reached. This means that the same data are repeated again and again. When saturation occurs, the researcher knows that there is no need to continue the study. To achieve this, the researcher enters into the research problem and investigates its full extent and dimension. In the design of grounded theory, the researcher collects data, analyzes it, identifies issues and develops categories to organize information conceptually. With these findings, returns to the field to gather more information on these topics and categories, analyzes data again and refines or expands the themes and categories in the light of new themes and categories that emerge. This process is repeated until information saturation is achieved in themes and categories. The themes that emerge from the data and the categories prepared by the researcher always guide the next step of data collection. Every effort to collect new data should facilitate the development and debugging of topics and categories (conceptual density). This is known as theoretical sampling or purposive selection of samples in order to understand the themes that emerge and develop categories of study to develop this grounded theory to explain the phenomenon under research. The goal of all research is to generate data that allows the practice of education to develop centered upon evidence. Scientific research helps generate and validate theories to guide practice tested and validated by data. The design of grounded theory is an inductive approach to theory development focused on data generated in the context of the study. Its strength lies in the systematic investigation of a phenomenon to the point of reaching data saturation. The product of this design are grounded theories, or diagrams, grounded in the study data. This is essential to generate theories and hypotheses as a basis for further studies.

e. **Research of social critique.** The objective of this design is to understand the political aspects of education: who has power, how it is transmitted or what structures are used. Education systems are political to the extent that they are consistent with the interests of those who organize and regulate them. The political influence on education can come in the form of programs, projects, textbooks, research funding, curricula or standards where those in power determine what is to be taught in the education systems. By doing so, impose their values and beliefs to the constituents of the schools. Historically, this has been the subject of tension and disputes between sectors of education because the recipients of administrative decisions do not participate in the decisions imposed. This is seen as an oppressive and undemocratic processes. The investigation of social critique seeks out the consequences that come attached to administrative, hierarchical and authoritative decisions that occur in all education

systems. Its aim is to improve education by studying projects, administrative decisions or school reorganizations to question the order, scope,
values, beliefs or effectiveness of these. In this research the voices of those
living the consequences of the administrative decisions are exposed. The
investigation of social critique seeks justice for individuals, groups or
communities that are affected by policy decisions of educational systems.

The field of education has a strong political undertone because it
responds to authorities that regulate and direct. Political influence is seen
in administrative, managerial decisions, design of intervention programs
or research, allocation of funds, the appointment of managers and even
the administrative structure that an educational system adopts. The most
common technique for research in the design of social critique is the analysis of texts (Merriam & Associates, 2002). Text analysis has been used
extensively in the field of education to discuss books, special programs,
regulations, orders, and institutional policies and school reorganisations.
Through this analysis values, beliefs, assumptions, ideals and intentions
that guide education are identified. This makes it easy to track how this
translates to the practice of education and its impacts. If the impacts of
projects or programs are understood, informed decisions about how to
improve them can be made.

2.6 Selection of Participants or Sample Study Criteria

The term "sample" is used in educational research to describe the process
of identifying and selecting appropriate information sources from which the
researcher collected data on educational phenomena studied. Most qualitative
research in education uses people as sources of information. Some researchers
use the term **study participants** rather than samples. Due to the social nature of
education, programs, institutional processes, curricula, instructional practices,
schools or documents are also studied. Some researchers use the term **units
of analysis** to describe those sources of information that are not people.
The unit of analysis facilitates identification of people who can help the
researcher to understand the phenomenon being studied. For example, say that
a successful sports program in a school becomes the phenomenon of research
study. Everything about this program such as its activities, policies and people
management and implementation, become possible sources of information to
understand its success.

The sample in qualitative research is always the selection of participants or
units of analysis where the research phenomenon manifests. For example, let's

say the research is the phenomenon of dropouts. Study participants are school dropouts. This is known as a **criterion sample**. The selected criterion means that only those people (participants) or units of analysis where the phenomenon under study is manifested are chosen. For example, if the phenomenon of research is the dropout, the **criterion sample** for the study are school dropouts. If the phenomenon under research is criminal activities in student dropouts, then the sample must include student dropouts with criminal records.

The criteria for selecting the sample are always the manifestation of the phenomenon under study. Criterion samples are based on the assumption that there is a **theoretical population**. A theoretical population is one in which the phenomena manifests what the qualitative researcher wants to study. The theoretical population represents the "theoretical" universe of possible cases where the phenomenon is manifested and where information can be gathered on it. For example, say that the phenomenon of study is the dropout of freshmen in universities. The theoretical study population would be all students who entered college in the country and dropped out in the first academic year.

The selection of the sample in qualitative research is always inductive. This means that after identifying the criteria to be considered, the sample builds and develops case by case, in the process of research. The aim of qualitative research is to understand and provide a thorough and detailed description or explanation of the educational phenomena under investigation. To achieve such richness and depth of qualitative data, it is prefered to work with small samples, case by case, to achieve this objective. For example, back to the dropout of students in their first year of college. Say the researcher identified five dropout students from different universities of the country for interviews. Three of these five interviews provided rich and detailed information and two were non-informative. Although the researcher has five interviews, there is the need to identify other dropouts to continue interviewing and try to understand the phenomenon of dropping out of college in the first year of study. In this process, the researcher will encounter descriptive interviews and very little other information. Understanding the phenomenon of dropouts is constructed case by case. That is why the qualitative sampling is inductive. The point of sampling is the richness of information that is obtained from the phenomenon under study and not the number of participants. As the aim of qualitative research is to understand, describe, or explain in detail the phenomenon studied (dropouts in the first year of university), **representativeness** emerges in the process of research. Representativeness implies how rich, descriptive and illustrative the data collected is from the phenomenon under study. If we go back to the example above, the researcher conducted five interviews, but only three of them provided rich and detailed information. The representation in this example lies in being able to describe in detail the phenomenon of college dropouts in the first year of study. A greater richness of information, a greater confidence of the researcher that is capturing the phenomenon studied. Although the sample selection began with the careful selection of participants where the phenomenon occurs, it is in the process of research one discovers how informative are the cases that were selected. That is why the representativeness of the qualitative sample is reached in the process of conducting the study. Supporting this process is the researcher's ability to interview and observe to penetrate the phenomenon studied and generate the detailed and thorough description pursued.

As the essence of the sampling process lies in the richness of information generated on the phenomenon under investigation, the number of cases or the sample size become secondary for these purposes. Usually, the richness of information is not related to the number of people participating in the study. For example, learning theory developed by the psychologist Jean Piaget was based on observations of his three children. From observing his children's learning he generated a detailed explanation of this process. The specificity of the explanation generated on learning facilitates appreciation of the same phenomenon in other children. In this example, three cases were sufficient to explain the phenomenon in detail. The sample size is determined during the process of research. **Two criteria determine the sample size: the richness and depth of data and if information overload is experienced in the study.** In some studies, these criteria may lead to 10 people and in others a 100. In qualitative research various types of samples are recognized as research needs and circumstances that arise in the course of the study:

a. **Snowball sampling.** The initial sample at the beginning of every qualitative study is of criterion or intentional. It is common for the researcher to begin a study with a specific number of participants, say 10, as the initial sample criteria, and this will increase in the research process because the participants themselves tell the researcher of the presence of other possible candidates for the study. The researcher may contact these potential participants to continue deeper into the phenomenon being studied. When this happens, we say that the sample was a snowball. In other words, the sample increases because study participants alert the researcher to other potential key candidates that do not necessarily belong to the initial theoretical study population.

b. **Discrepant or negative case samples.** There are occasions where the researcher can deepen the understanding of a phenomenon by studying the **negative cases**. A negative case is a person/analysis unit where the phenomenon is expressed in reverse. In the example in paragraph 3, college students who dropped out in their first year of study were selected. A study using negative cases would take students who have successfully completed college. Sometimes, when considering the negative or opposed cases, the researcher can better understand the phenomenon being studied.

c. **Perspectives and emerging theories samples.** A third alternative of sampling is the research of emerging explanations. In other words, the researcher begins the investigation and in the course of the study different perspectives emerge on the research topic that merit investigation. Venturing into these perspectives will lead to other persons that were not considered in the initial criterion sample of the study. Following these emerging explanations help to better understand the problem and refine it because it has more than one possible explanation of the phenomenon.

2.7 Data Collection

Data collection is the organized process that is carried out to obtain information about the phenomenon under research. The goal of this process is to move from one level of knowledge to another level of knowledge about the research problem. As phase of an investigation, data collection is planned and carried out taking into account the requirements of the scientific method or the quality standards of the research model (Ander-Egg, 2003.) (a) Must be systematic and organized, expressed as a set of operations whose purpose is to obtain the estimated information necessary to collect. (b) With clear and deliberate explanation of its purposes. (c) Ensuring sufficient guarantees of reliability and validity. (d) In each case choosing those most relevant techniques and procedures in accordance with the nature of "that" to be studied and the type of information to be collected. The methods of qualitative research lie in the interview, observation and document analysis as data collection techniques. The techniques are not created independently of its methods. These are the tools to cause the act of research. The intent of the techniques is given by the philosophical, ideological and ethical reasons of the method (Ander-Egg, 2003). These reasons should be observed in processeses aligned to their purpose of entering these phenomena being studied. The purpose of the interview, observation and document analysis is to help the qualitative researcher capture the phenomena studied in educational institutions. Using these techniques must allow exploring, entering into, understanding, describing and explaining the dimensions of the educational phenomena being studied: how they are, where they come from or what their symptoms are. In qualitative research, the effectiveness of the techniques, and data collection, comes in the form of deep understanding of the phenomenon that is generated with the data collection, the ability to describe in detail the phenomenon and the density or richness of the data collected.

d. **The qualitative interview.** Is a conversation between two or more people to deal with an issue (Ander-Egg, 2003). As a research technique, the goal is to collect information on the research problem. The interview can be from person to person (individual) or group (focus group interview). It is used in those studies where is is not possible to observe the behaviors or feelings of people who live or experience the educational phenomenon being studied. This is common in subjects with phenomena, situations or events that have already occurred. An example would be interviewing defectors who dropped out. It is also used for the study of future events. For example, what would you do with your studies if economic conditions do not improve in your family? Its development is focused on the topic and the research. For the researcher, the interest of the conversation will be the research problem and for the participants will be sharing their inner experiences with the phenomenon. The dynamics of qualitative interview focuses on a dialogue of questions and answers: the researcher formulates questions and the participant answers them in their own words. There are no right or wrong answers because the conversation revolves around the experiences, emotions, values, culture and attitudes of the participants to the phenomenon/problem being studied. Through interviews, the researcher knows the research topic from the perspective of those who live or experienced it. Typically in qualitative interviews the conversation begins with a theme and goes to another and then returns to the original topic. It is not always that the development of the qualitative interview is linear and follows the order of the questions set. Each participant is unique. Some interviews will be richer in information than others. With some participants, the conversation will flow easily because they are more outgoing than others. Its development must occur in a friendly, cordial, pleasant and respectful dialogue producing empathy between researcher and the participant. If the participant does not feel comfortable with the dynamics they will probably not express themselves openly and in confidence. This may produce unreliable information.

1. **Types of interviews.** The two most common types of qualitative interviews in education are:

   a. **The in-depth interview.** This type of interview is focused on in-depth understanding, opinions, attitudes, experiences and personality of the study participants or groups. Its purpose is to understand the phenomena, situations or experiences that are studied from the perspective of the participants. It is an open interview where the researcher explores the research topic in its entirety and dimension. This allows exploring anything else that emerges unexpectedly in the conversation, but that is relevant to the study. As a research technique, it entails the researcher being knowledgeable on the subject in order to ask relevant and probing questions. It also entails experience as an interviewer to ask questions tactfully, but with the ability to motivate the interviewee to elaborate and penetrate into their experiences or events addressed. It involves the researcher's ability to hear the details, organize the information communicated by the respondent and to integrate it into his/her mind to create a clear picture of the participant's perspective.

   b. **The focused interview.** Is a semi-structured interview that aims to understand situations, experiences or specific events that individuals or groups have experienced commonly. For example, the process of enrollment in a university or training program of faculty in an educational institution. The researcher previously studied the situation, the experience or event and formulates questions about the specific aspects they want to know. In light of these questions, develops focused interviews to try and learn about the phenomenological aspects of the experience: cognitive, evaluative or emotional reactions of individuals that live it. In this type of interview, the interviewee expresses freely and in their own words. The interviewer will seek deep, concrete and specific responses to the event or situation being studied. This implies experience, skill and tact of the researcher.

2. **The role of the researcher in the interviewing process.** The aim of qualitative research interviewing is to understand the phenomenon being investigated, from the perspective of those who live and experience it. To accomplish this, the role of the researcher is:

To ask open-ended questions to enable the participant to speak freely, listen carefully, be empathetic, not comment or pass judgment on the positions assumed by the participants, gather examples and clarification when they don't understand and collating this understanding with the participant to verify the information.

3. **General procedures for qualitative interviews.** The interview process goes through the following phases and activities:

    a. **Design the interview to be held.** The interview is designed in the light of the problem and the research questions of the study. The interview format is organized by questions/topics that facilitate a conversation starter (icebreaker), develop dialogue and conclude (having a closure). Both the researcher and the interviewee should conclude the interview with the feeling that there was a beginning and an end to the conversation. It is prudent to conclude the interview by asking the participant if there is something they wish to express that was not discussed in the interview. It is also important to ask the interviewee if they have any questions they want to make regarding the interview or study. It is prudent to conclude the conversation leaving the door open in the event of requiring a second interview to delve into issues or to validate the information once it is transcribed. The study and the interview should be transparent to participants or interviewees. Two techniques dominate the design and implementation of the interview:

        i. **Interview guide.** A common technique for qualitative interviews in education is to develop an interview guide. An interview guide is to capture on paper the questions or issues that are needed to be discussed with the interviewees. These questions are planned, according to the structure of the interview: Questions for the initiation, development and completion of the conversation. The function of the guide is to give order and sequence to the interview, assisting the researcher to ensure covering all the topics needed to be investigated and use time efficiently. Interview guides can be evaluated by colleagues to verify and validate their vocabulary, the order of the questions, the questions regarding the purpose of the study or their clearness. They can be practiced prior to the research to study the reactions of the respondents, meet potential reactions to questions and rehearse interview technique.

   ii. Protocols for the interview. Another common technique to accompany the interview guide is the use of protocols. Protocols consist of writing on paper, word for word, the instructions each study participant will be given and step by step details to be followed by the researcher when interviewing. These protocols can be evaluated by colleagues to validate vocabulary, procedures, clearness, specificity or relevance of the procedures before, during and after the interviews.

b. Practice the interview before the study. This will facilitate a better appreciation of the kind of situations which may be encountered when formally making the interviews. Although in qualitative research, the researcher can clarify the question to the respondent, the rehearsing of the interview can help better understand the relevance or complexity of questions for potential study participants.

c. Document the interviews. It is common to use audio recorders to document the conversation and use notebooks to record comments, ideas or expressions relevant for the study, and that emerge in the interview, that can not be captured in recording. The use of recording (audio or video) and notepads needs to be discussed with participants prior to the interviews so as not to surprise them, make them feel uncomfortable or intimidated. Making notes during the interview can disrupt the dynamics and chemistry of the conversation, cut the inspiration of the interviewee or instilling distrust interviewed by making notes upon only some aspects when everything can be relevant to the participant. This aspect of the interview should be carefully planned and whether you should be doing this during or after the interview.

d. Convert the interview into data. Interviewing without studying the information collected will not help much in a study. Two techniques are used to convert interviews into qualitative data:

   i. Transcribe. Transcribing the interview is a common strategy to initiate the process of analyzing the information and turn it into data. The transcript has the advantage of allowing the researcher to read and check the information that is relevant to their research goals. Not all information collected in interviews is always valuable. The transcript can be electronic or paper based. Once transcribed, the researcher does not need to listen repeatedly to recording interviews to study them.

ii. **Analytical memos.** This technique involves writing notes on the transcripts that are made in order to highlight and provide what information they have and that what is lacking in the data to understand the phenomenon or answer the questions of the study. The memos are analytic reflections the researcher made in performing the interviews to monitor their data and remain aware of the information they have.

4. **Philosophical and methodological alignment of the interview as a qualitative research technique**. There are many interview techniques, some very structured and others unstructured. For the interview to be considered a qualitative research technique, its procedures must reflect the following aligments with the model of qualitative research in education: First, the respondent can express themselves freely, in their own words, build their response and express their experience with the phenomenon under investigation. Second, the researcher can investigate that experience. If the interview does not provide for these two criteria occur, then there is no space in the dialogue for the phenomenology of experience emerge and be explored. In this technological age, creativity in communications continues to evolve. Video conferences, emails, conference calls and video recordings videos are shortening the physical distance between people, opening new communication gaps and shortening the time to obtain data. We acknowledge that technology will continue to redefine the interview as qualitative research technique. In some cases, we agree that it enables the execution of certain interview studies. This should not mean that the "convenience" of interviewing people by electronic media replaces the face to face interview when there is no methodological justification.

e. **Observation.** Educational institutions manifest an infinity of dynamics, situations, events and phenomena that the researcher can witness and experience with their own eyes and senses. Observation is a common technique of qualitative research. It is considered a direct technique for data collection. As a research technique, transcends the mere fact of looking at things or making a record of facts and events (Ander-Egg, 2003). Observation is a technique where sight, sound and conversation are used to collect data. Observing is not an isolated activity. It involves listening and sometimes asking to ascertain what is observed. It is used to collect data and verify these. Almost always it is used to study situations as they occur in the social life of individuals or educational institutions. You can observe events, conditions or behaviors of people present in their natural environment. Common aspects that are studied through observation are the physical environment of the school, program development, interactions generated by the curricula, conversations that occur in leisure, clothing, oral and body language and behaviors taught in school and are manifested in the home or community. It is also used to study phenomena and behavior in preschool children and students with special needs who can not express themselves clearly verbally.

1. **Types of observations.** The following two observations techniques are common in educational research:

   a. **Participant observer.** Meaning that the researcher engages with people and activities, events or situations being studied to experience firsthand what the rest of the group is living and experiencing. The researcher is not merely a passive observer and what happens is that they become one of the group. "Putting on the shoes of others and walking a mile in them" provides that understanding which is needed when the researcher is studying unfamiliar phenomena. The participant observer technique has been used in education in many ways: to understand the culture of a school from the perspective of students, to enter the values of sports teams, experience life in disadvantaged communities or criminal cultures, to experience what it means to grow up without parents or in foster homes. As a research technique, and for data collection, it is complicated and time consuming, also requires a great amount of physical, mental and emotional energy. It involves empathy, compassion, and respect for the culture and values of others.

b. **Non participating observer.** Means that the researcher actively observes and study the phenomenon, but without becoming a participant. Observing a phenomenon actively means carefully observing the unfolding of events (which seeks to understand the dynamics), what happens, and validate what is seen. Non participant observation can take two forms: open or incognito. In the open observation, the observed are aware that the researcher is making observations while the incognito observation, they do not. This point is debated in some forums on ethics and responsible conduct in research.

2. **The role of the researcher in the process of observing**. The aim of the researcher observing is to gain a real understanding of the phenomenon studied to describe it in detail. Typical themes in education are observing the environment/physical setting where the phenomenon occurs, activities/dynamics that occur there, people who participate in the activities and what they mean to them. By collecting data through observations, the researcher has to be aware that the reader of the study must understand the phenomenon without being present. The observational data should provide the reader with the same understanding that the researcher achieved: what happened, why and how. The description of the observations must be factual, accurate and meaningful. In other words, free of minutiae and irrelevancies that do not add to the understanding of the phenomenon under study. The role of the researcher observing is to take descriptive notes of what is observed without becoming overwhelmed in detail, distinguish the essential from the trivial in what is observed and described. It also involves seeking reliable ways to verify what is seen.

3. **General methods of observation as a data collection technique.** The effectiveness of observation as a data collection technique rests with the researcher as research instrument. The procedures, standards and practices to regulate observation as data collection technique focus on making this a reliable technique. The following recommendations serve as two important considerations of observation: the methodology used and the behavior of the researcher (Ander-Egg, 2003).

a. **Clearly define the research objectives.** Observation is most effective when there are defined research objectives that establish clearly what information is needed and why. For example, in ethnographic studies, where the aim is to study the culture, values or beliefs of the groups, the observed elements are defined with the research design. This reference framework serves to focus the observations, to interpret the data and to contextualize the findings. This contributes to the accuracy of the observation if the role of technology in the design of the study is defined. For example, the accuracy of the technique increases if the role of the observation is exploring or describing an event. If the goal is to explore, the focus of research is defined and refined in the process whereas if the objective is to describe, the focus is defined from the start of the study.

b. **Identify tools to support observation.** Observing is work involving concentration and dedication. The researcher should seek to be responsive to the observation design planned so as not to be distracted in data collection and deviate from their goal. Three strategies that increase the effectiveness of the observation are as follows:

   i. **Observation guide.** An observation guide is a list of topics, questions or issues that is wanted to be observed. The guide provides direction to research and helps staying focused on data collection and analysis. Some researchers use videorecording to register the phenomena without having to be physically in the educational scenario. Afterwards, study the video as often as necessary. For example, a friend used observation in his dissertation. Video recorded and analyzed classes at home using their observation guide. This made it easier to question and reflect on what was observed.

   ii. **Records and schedule of observations.** Observing without documenting will not allow much progress in the development of research. Keeping a record of observations is another strategy that increases the effectiveness of the technique and the reliability of the data. Documenting observations allows studying the notes and reflection on these. Some researchers make their notes in notebooks. The time of documenting the observation contributes to the effectiveness of the technique. Some researchers make notes while watching. Others make notes after completing their observations. The chance of forget aspects and details increases if the entries occur after the researcher has concluded their observations.

c. **Reflective journals or analytical memos.** Documenting without analyzing notes, also will not facilitate progress in the development of the study. Annotations should be regularly evaluated to turn these to data. This exercise consists of three aspects: clarify the information gained, identify what information is missing and needs to be collected in the following sections of observation, and identify which areas are still unclear and more information is needed to clarify them. This analysis exercise is recorded on summary sheets that summarize the state of affairs/development of the study. **Make more observations in the light of the reflective journals or analytical memos.** Redefine observations in the light of the data needed to complete the study.

d. **Validate observations.** Observation as a research technique is not merely to watch and describe things that happen. This is the descriptive part of the observation. The interpretation of what is observed has to be validated. In some cases, the validation will support the interpretation of the dialogue with participants. In other cases, validation is the ability to observe a behavior or manifestation of a phenomenon repeatedly to the point of being able to explain why and how. This is the case when studying infants and children whose verbal development is limited.

e. **Maintain ethical conduct at all times.** When entering an educational institution, the researcher will follow the communication channels or search for the contacts assigned. These are the resources that will introduce the researcher to the people or groups to be observed. The formal presentation of the researcher in an educational institution is essential to establish goodwill and to win the trust of the participants. At the presentation of the observation scenario for the first time, it is proper to explain to the participants the purpose of the study and the activities conducted by the researcher. This helps to clarify doubts, suspicions and overcomes the phenomenon of strangers on school grounds. In the following observation visits, the researcher will be integrated as discreetly as possible so as not to disrupt or affect the daily development of what they observe and study. Similarly, if there is a need to approach a participant to discuss and validate an observation, it can be done in the most natural way possible so as not to cause misunderstanding in the participant or the rest of the group observed. Always perform observations in the days and hours assigned or permitted as agreed.

4. **Philosophical and methodological alignment of observation as a qualitative research technique.** Observation, by nature, is descriptive and generates qualitative data. These two criteria are not sufficient to classify a study in the category of qualitative research. Observation becomes a qualitative research technique to the extent that allows the phenomenology of human behavior to be observed and validated. Describing the physical condition of a school does not capture the phenomenology of human behavior. Describing the behavior, emotions and vocabulary of children in the context of a class in a dilapidated room is closer to the nature of qualitative research. Corroborating this observation guarantees the validity of the data and authenticates the phenomenology of the experience. That procedure raises the observation to the status of scientific research. In this information age, technology has redefined the observation technique in qualitative research. Video recorders on cell phones, cameras on computers, tablets, digital video cameras of reduced size and many other electronic devices allow electronic observations in places where it was once not possible. The element of privacy and research ethics are also redefined with very creative possibilities enabled by technology. Technology should be a tool for the researcher and not the means of investigation for convenience displacing observation as a qualitative research technique. Data quality and the possibility of testing must be criteria of weight to select electronic medium to collect data through observation.

f. **Document analysis.** Document analysis is another valuable technique for data collection in qualitative research in education. Highlighted in some books and not others. The analysis of documents, when available, contribute in many ways: To understand educational phenomena, situations and study programs, to track the impact of administrative decisions and refine the research design and develop more relevant research questions. A document can be material written, visual or an artifact. Documents can be official reports of the educational system, circular letters, personal documents, online documents, books, photos, videos and any mechanism that provides information on the research problem. In the field of education many documents that allow approach to research these phenomena are generated. Documents in education facilitating the understanding of administrative aspects of education can be memos, academic standards, student handbooks, teacher handbooks, educational vision statements, goals and objectives, philosophical positions, labor laws applicable to schools, and teacher evaluation systems and accountability of the education system. Learning can be understood through the analysis of school curricula, textbooks, academic standards and standardized tests that define academic areas. These documents also allow appreciation of the teaching techniques and expectations that must be met by the educator to achieve them.

Document analysis is considered a technique to collect complimentary data. The first challenge in using documents is to establish their validity: if they are authentic documents of the educational system, if the information provided is accurate and useful for the purposes of the study and whether it is a primary (directly from the experience of the phenomenon) or secondary information (information from sources not directly involved with the phenomenon). The second challenge in using documents is the ethical element of identity protection and safety of participants that emerges from these. In this era of copyright, the researcher should be well aware of this fact when using and citing documents. Document analysis follows the following procedure:

1. **Document selection.** The main criterion in selecting a document to use as data collection technique is content relevance to the study. There is no magic formula beyond the common sense of the researcher to relate the content to their research. Aspects that help to evaluate the document, beyond the content, are the date of publication, the purpose of publication, situation or context it addresses and the professional nature and status of the author.

2. **Uses of documents.** The researcher may encounter documents before, during and after completion of their study. Depending on the timing, documents can serve several purposes: **Encountering documents before data collection** can help to better understand the theme, define the research design or in writing more relevant research questions. It is common in qualitative research encounter

documents **during data collection**. When this is the case, the contents of the document can help to better understand situations, controversial or circumstancial surrounding the research problem. This in turn can serve to redefine the design of the study, the research questions or develop new questions/hypotheses. Document information can serve to corroborate some of the data already collected and increase the validity of the research design and self-esteem of the researcher. **Encountering documents after the study** may help the researcher deepen their findings. The content of the document can be compared with the findings for more elaboration or tune them further. The information can help identify emerging areas of research and propose new questions.

3. **Philosophical and methodological alignment of document analysis as a qualitative research technique**. A lot of information can be generated from the analysis of documents. To the extent that the content of the document allows understanding the phenomenology or the reasons, emotions, feelings, attitudes or contexts of situations, events or programs, the greater its potential as a qualitative research technique. When a document has the phenomenology of its protagonists, the researcher can locate this in temporal contexts and try to track them over time or relate it to historical moments to explain philosophical positions, visions or decisions that are unseen, but determine programs and educational decisions.

2.8 Analysis of Qualitative Data

Analyzing qualitative data means to identify and interpret the messages they contain. It is a process that involves studying, understanding and organizing the information gathered in the study to extract their meaning. Although there are programs and technologies that aid the researcher in this process, none replace the mental exercise that involves studying, analyzing and interpreting data. In general, analyzing qualitative data is a process that includes the following phases:

Phase 1: Understand the data. The data analysis process begins with understanding the study information. This means **knowing its contents or messages**. The data in qualitative research consist of all the information that is collected from the study. This information comes in the form of words, ideas, phrases, diagrams, charts, drawings or photos. Good qualitative data is rich in descriptions of the educational phenomenon being studied. The more descriptive information is upon the phenomenon that is studied, it becomes easier to understand. Qualitative data describes aspects of education such as: (a) **Behaviors of individuals or groups.** Examples of these may be specific behaviors, interactions, relationships, emotions, values, beliefs and cultures, (b) **Structures of the academic educational institution.** An example of this can be educational programs, curricula, management structure or processes that operationalize the functioning of the institution, and (c) **Interactions of individuals/groups with the structures of the educational insitution.** For example, situations that occur due to compliance with institutional rules, reactions to physical properties of the structure such as classrooms, or the perception of safety on campus. In qualitative research data can be understood in the research process or in the final data analysis.

   a. **Knowing the data in the research process.** This is possible if the researcher pays attention to information collected while directing interviews or observations. This understanding is deepened if analytical memos (interviews) or daily reflectivions (observations) are used as techniques to understand and debug the data as the study is conducted. Some researchers analyze interviews or observations immediately upon collection. In this way they familiarize themselves with the information whilst collected.
   b. **Understanding the data as a result of the data analysis.** Means that the researcher is studying data after collection. The research focuses on collecting all the study information and then the researcher dedicates themselves to the analysis and interpretation of data. This is apparent in studies of interviews. There are occasions where the researcher manages to coordinate interviews consecutively and this does not allow sufficient time to analyze data immediately upon collection. This may occur when trying to perform a qualitative study in a short time.

The first formal step to understand the information collected is to organize this material and turn it into a useful resource for data analysis. For example, an interview of one hour can be translated into a transcript of 20 double-spaced pages. If the sample of the study is 30 participants, we are talking about 600 pages of transcripts. If the researcher does not transcribe, then they would have to listen to recordings countless times to extract the relevant information they contain. Some researchers prefer to transcribe the information and others choose to write summaries. Transcribing the information or not is a decision of the researcher to respond to how that decision facilitates their ability to effectively analyze data. Transcribing the information is a requirement for using available computer programs for analyzing qualitative data. Here are some tips for managing transcripts, whether they are paper or electronic: (1) In interview studies, generate a transcript for each participant so that you can study them by individuals and as a group. (2) In observational studies, transcribe by observational sessions. This way you can study the gradual approach to the phenomenon. (3) Enumerate the pages and lines of the transcripts. This will facilitate access to the pages and lines where the relevant information of the phenomenon is studied.

Phase 2: Identify themes: It means read the transcripts to know their content and identify themes, patterns, or perspectives they contain about the phenomenon being reseached. **A theme is an issue that is manifested or expressed in the data about the phenomenon being researched.** Its secondary if the transcript is paper or electronic. The point of the exercise is to become familiarized with the information to identify themes that can reduce the volume of data. Two analyses proceed in this step (Miles & Huberman, 1995):

a. **Identify themes per study participant.** The researcher proceeds to study their transcripts, interview by interview or observation by observation. The goal is to identify themes per participant/observation. Consider the following example in the study of the motivations of children to exercise:

> **Research problem.** Motivations for children to exercise
> **Interview Question:** What motivates you to do exercise?
> **Answer of one of the participants:** I like to be fit and spend time with friends.
> **Example.** Summary Table of Participant 1 interview

Question	Answer summary	Interview page
1	Physical efficiency	Page 1, Paragraph 2
2	Like to sweat	Page 3, Paragraph 1
3	Prefer exercising in a group	Page 3, Paragraph 6

In this example, the researcher summarized in a table all interview responses and identified the page and paragraph where to locate the themes. This makes it easy to locate information in the transcripts and identify segments or verbatim to support the finding. This exercise of initial data analysis proceeds between interviews or between observations.

b. **Identify common themes for the group of participants.** Means that the researcher proceeds to study their transcripts in a transversal manner. This implies seeking to identify common themes among participants. Consider the following example using the study of motivation for exercise in children:

> **Example.** Summary table of transversal topics for interview questions

Question	Answer summary	Interview page
1	Physical efficiency	Page 2, Paragraph 3
2	Physical efficiency	Page 1, Paragraph 1
3	Recreation	Page 2, Paragraph 3

In this example, the researcher summarizes the responses of all participants in a table, by question, and this allows to see the most recurrent transversal topics (eg., physical efficiency) and less recurrent (eg, recreation).

Phase 3: Develop categories to organize and explain the themes. The themes identified in step 2, are used to develop categories that allow organization of information, summarizing, interpretation and verification. **A category is a classification that is assigned to locate themes in groups or their peculiarities.** The challenge is to identify the feature or pattern of the theme to assign names to the categories. The titles, or names of the categories, must accurately describe the specifics of the topics. Categories must meet two requirements: accurately describe the themes and understand them without a lot of explanation. Consider the following example of a phenomenological study of physical fitness in children aged 10 to 15 years:

> **Interview Question:** What motivates you to exercise?
> **Child's response:** I like to be fit and spend time (hang out) with friends.
> **Possible categories:** Physical efficiency, socialization, recreation.
> **Expanation:** The category of physical efficiency is apparent from the first part of the answer, "I like to be fit." The categories of socialization and recreation emanate from the second part of the answer, "to spend time (hang out) with friends." The category

of physical efficiency describes the child's motivation to exercise to stay fit. The categories of socialization and recreation describe the child's motivation to exercise to socialize and be with friends.

Identifying themes and developing categories helps the researcher to organize the information for their particular features or issues. This allows knowing the information held for each category and determine whether or not this is illustrative. Two exercises can help researchers refine their categories:

a. **Summarize the categories in tables.** The objective of this exercise is to visually appreciate the information you have. An example to study information about exercise motivation follows. You may notice that the researcher organized the response into three categories: Physical Efficiency, Socialization and Recreation. The table summarizes the categories and themes that they generate. The categories and topics are linked to the participants in case you need to revisit the transcripts to evaluate the study data and identify segments of the interviews to support them.

Example. Summary table of the categories from the question of motivations for exercise

Categories	Themes	Participant
Physical efficiency	Physical appearance	1, 3, 7
Socialization	Spend time with friends	2, 3, 4
Recreation	Amusement	8, 9

b. **Quantify.** This technique consists of identifying patterns in data and making the use of frequencies and percentages to reduce the information and describe the pattern. The goal to quantify is not to measure, but to produce a numerical description of the pattern as a technique for summarizing data and corroborate it. The technique of quantifying has the advantage that it becomes an effective strategy to corroborate data and this increases the validity and reliability of the interpretation by the researcher. It has the disadvantage that it is not applicable to qualitative studies using small samples. Consider the following example:

> **Interview Question:** What motivates you to exercise?
> **Child's response:** I like to be fit and spend time (hang out) with friends.

Possible categories: Physical efficiency, socialization, recreation.

Application of the technique:

Category	F	%
Physical efficiency	5	50
Recreation	3	30
Socialization	2	20

N = 10 participants

Phase 4: Corroboration of categories. Corroboration means collating correspondence and accuracy between categories, their content and the interpretation made of these. The fundamental question from the researcher here is whether the data and the categories are in support of such interpretation. Corroboration establishes the validity and reliability of the analysis. The following procedure is typical to corroborate the validity of the categories that are developed:

a. **Evaluate categories.** Evaluate categories and definitions used to create them, in light of the issues explained. The central point of this step is to determine the correspondence between the subjects (data) and categories. This exercise involves reflection and evaluation of the categories and how they accurately describe the themes and information they represent. If there are aspects of the themes that the category has not completely covered, the the category name or themes must be redefined to achieve the consistency between themes and categories.

b. **Develop an explanation of findings.** Once the corroboration of categories is complete, the researcher is in a position to develop an explanation of their findings. This means explaining how the data collected and categories are organized to meet the objective of the study and answer the research questions. Depending on the volume of information, a common strategy at this point is to use the research questions of the study to present data as sections of the findings chapter. Information that can not be articulated in these schemes can be presented as supplementary material in the appendices of the study.

c. **Evaluate whether it is necessary to collect additional data to document explanation of findings which warrants developing a grounded theory to conclude the study.** Many qualitative studies in education end at the step of **evaluating categories**, because its aim is to describe a phenomenon. If it is necessary for the study to explain a causal relationship found, or if the aspiration is to develop a theory that explains the data, and does not have it, it will be necessary to return to the field to conduct further interviews or observations to gather the information needed. The need for additional information is presented when the researcher can infer the possible explanation or connection between their categories, but can not produce evidence with the data compiled.

Phase 5: Presentation of findings. The presentation of findings involves two considerations: written report content and style of communication. Common styles of presenting the findings of qualitative studies are:

a. **Narrative.** The narrative is a common style of presenting the findings in qualitative research. In this style, the researcher addresses the reader in the first person and "tells" the study's findings. Many books refer to the narrative as "the story" about the findings. The effectiveness of the narrative rests on the researcher's writing style and his ability to present findings in an organized and balanced manner. Organized means in an orderly manner and in sequence. Balanced means that the narrative allows the reader to identify and differentiate the voices of the study participants from the interpretation of the researcher. The narrative consists of a combination of explanations, anecdotes, examples, concepts, categories and interpretations using the raw data of the study. For example, display some verbatim quotes from the interviews to support the finding presented. The following example illustrates this effect:

> Physical efficiency was the motivation that most repeated among study participants. This category was titled with the name of physical efficiency because the reasons or motivations, expressed by the study participants, to get involved in exercise programs are related to fitness and health. This is known in the literature as physical efficiency. In this category, five reasons to exercise are collected; weight loss, bodybuilding or develop a more muscular physique, develop greater cardiovascular endurance, flexibility or strength to compete in their favorite sports, "to be in form," feel good about themselves and the

image they have of their body. So said one of the participants ... "I like to be fit and spend time with friends" (Interview 5, p. 17).

b. Combining narrative with visuals. As qualitative data consist of words that describe these phenomena under reserach, to present and discuss the findings in narrative form can be a complicated task for the researcher and the reader of the study. Some qualitative researchers resort to using tables, diagrams or matrix to enrich narrated presentation of their findings. In this technique, the researcher presents the categories in tables or diagrams and then explains them. The tables and diagrams enrich the presentation of findings by summarizing the data visually, allowing quantification, if necessary, and helps the reader to check their own interpretations with the data presented by the researcher. The visual narrative consists of tables where the categories are presented and the findings are summarized by frequencies and percentages. Then the researcher explains the category and its contents. The following example illustrates this style:

Table 2.1 summarizes the motivations for exercise.

The dominant motivation to exercise was being fit or physical efficiency. This category was titled with the name of physical efficiency because the reasons or motivations expressed by the study participants to get involved in exercise programs are related to fitness and health. This is known in the literature as physical efficiency. In this category, five reasons to exercise are collected; weight loss, bodybuilding or develop a more muscular physique, develop greater cardiovascular endurance, flexibility or strength to compete in their favorite sports, "to be in form," feel good about themselves and the image they have of their body. So said one of the participants ... "I like to be fit and spend time with friends" (Interview 5, p. 17).

Table 2.1   Motivations for Exercise

Category	F	%
Physical efficiency	5	50
Recreation	3	30
Socialization	2	20
N = 10 participants		

2.9 Validate the Data

Qualitative research focuses on studying the social phenomena of education. To study these phenomena, the researcher uses the experiences of those who live and experience them. So, how is the data validity of those phenomena that only exist in the social world of education systems established? In educational research the terms internal and external validity are used to assess the accuracy of the data.

a. Internal validity. Means the correspondence between the study data with the phenomenon or problem investigated. The central question of internal validity rests on whether the researcher managed to faithfully capture the educational phenomenon investigated. Establishing the validity of the data focuses on two requirements that the qualitative researcher must satisfy: to establish the accuracy of their data and matching these with the phenomenon being studied (Creswell, 2007). To achieve valid and accurate data, the researcher must first demonstrate that they understood the experience of the participant and second that they accurately captured the phenomenon investigated. There are many papers in the literature on how to establish the internal validity of a qualitative study (eg, Creswell, 2007; Licthman, 2006; Mertens, 2005; Silverman, 2005). Let's examine the techniques for establishing the validity of the data.

1. **Prolonged contact.** Means that the researcher will interview or observe, for the time needed to understand, with certainty, the phenomena it studied. Certainty means that he has no doubts about the qualities, characteristics or manifestations of the phenomenon under study. The researcher collects sufficient data, has specific examples, detailed descriptions and observable indicators of the phenomenon. This data can be validated by second and third parties related to the phenomenon. For example, say that ethnography is done on the culture of poverty in the community where the school is located. As a result of prolonged contact in the community, the researcher is able to explain with clear examples, those behaviors of the culture, and beliefs that cause them, but that affect the value and importance ascribed to education students receive. He is also able to explain how these behaviors and beliefs affect the academic success or failure of students. Once the researcher explains these behaviors to the faculty and management of the school, they can relate to the situations and experiences they have had with students. These behaviors can be validated by members of the community from which the students come because they recognize them. The technique does not always involve

prolonged contact over time, but to achieve a significant interaction, deep or rich, with study participants and the phenomenon under investigation (Merriam, 2009; Creswell, 2007; Mertens, 2005; Lincoln & Guba, 1985).

2. **Information saturation.** Means that the researcher continues researching to the point that the data begins to repeat interview after interview or observation after observation. The saturation of information is the sign that there is no more information about the phenomenon under investigation. When this occurs, it is understood that there is no need to continue researching the topic because a saturation of information has occured (Merriam, 2009; Creswell, 2007; Mertens, 2005; Lincoln & Guba, 1985).

3. **Analysis of discrepant cases.** Means that the researcher is studying those cases opposed to the phenomenon being researched. For example, if the study is on school dropouts, discrepant cases are those students that under the same conditions of life, persevere in their studies and succeed. The study of discrepant cases can help researchers better understand a phenomenon by examining the opposite side. For example, at the time of elections it is common for candidates for governor debate the issues of the country. In these debates different positions can be appreciated on the same topics. The application of this technique would involve delving into each of these viewpoints to understand them and use them to explain the phenomenon under study. The logic of the discrepant cases technique is to understand a phenomenon by examining opposing manifestations (Merriam, 2009: Creswell, 2007: Mertens, 2005: Patton, 1990: Lincoln & Guba, 1985).

4. **Multiple researchers.** This technique is used in qualitative studies and field observations. The aim is that more than one researcher studies the same phenomenon to then see if the same behavior, emotions or events are observed. If two or more researchers observed the same thing, it is assumed that the same phenomenon has been observed and this is taken as evidence of a reliable approach. (Merriam, 2009; Creswell, 2007; Mertens, 2005; Patton, 1990; Lincoln & Guba, 1985). This technique is difficult to implement for people performing theses or dissertations, where the research work is individual.

5. **Differentiation of emic and epic perspectives of the study.** The objective of this technique is to provoke reflection on the positions identified in the data. The researcher questions whether they captured the perspectives of the participants or whether their interpretations as a researcher emanate from the information collected. This distinction

is important because the goal of the researcher is to accurately capture the phenomenon by researching the experiences of the participants. If these experiences are not accurate, then it is not possible to understand or capture the phenomenon studied through them. Many qualitative researchers produce reflective journals of their interviews or observations. In these reflective journals they record these interpretations, thoughts and ideas that emerge from the act of researching and collecting data. Thus, reflecting on the data collected. This helps to maintain separate perspectives of the participants of their interpretations. This technique increases the validity of the study to the extent that helps the researcher to distinguish between the experiences of participants and the characteristics of the phenomenon, and their positions as a researcher (Merriam, 2009; Creswell, 2007; Mertens, 2005; Lincoln & Guba, 1985).

6. **Corroboration of the participant.** This technique means that the researcher corroborated with the participants of their study the information gathered from these. For example, if it's an interview study, it is possible to show the transcript of the same to the interviewee for verification. In doing so, the participant can tell the researcher whether or not the transcript captured what they meant in the interview. The same logic applies if the study is observations. Transcription of the observation is shared so that the participant may validate the description or interpretation of the observation of the researcher. For example, the researcher observed joy because the child laughed a lot during the physical education class. The participant is approached and asked if they enjoyed the class and if they liked it. Corroboration with the participant helps the researcher to establish the veracity of the information collected because it captures the experience of the participants (Merriam, 2009; Creswell, 2007; Mertens, 2005; Lincoln & Guba, 1985).

7. **Internal consistency of the data.** The researcher uses the study data to corroborate postures, clarify manifestations of the phenomenon or challenge inconsistencies in the information. It can also means identifying patterns, trends or repeated positions in the data collected. When this occurs, the researcher can make inferences with complete certainty by the consistency of their data. For example, in an interview study on the implementation of a new education policy to increase student retention, the researcher identified that seven of the 10 impacted sixth grade teachers expressed dissatisfaction with the rules because they will have to complete more administrative reports. Although they understand what the policy pursues, they do not like the work that falls on them. Because

of this consistency in the data, the researcher can state with evidence the potential positive and negative impacts of the regulations. Consistency of data helps the researcher to pinpoint their interpretations that are supported by the data. In qualitative studies, the internal consistency of the data means the clear identification of the presence or feature of the phenomenon studied (Merriam, 2009; Creswell, 2007; Mertens, 2005; Patton, 1990; Lincoln & Guba, 1985).

8. **Triangulation techniques.** Means that the researcher uses more than one research technique in order to increase the validity and accuracy of the data. For example, in one study field observations and interviews as research techniques are used. This data is then contrasted with these techniques to determine the consistency and accuracy of information. Some researchers triangulate their data using official documents of the educational institution. Others resort to available studies on the same subject or positions documented in the literature to argue about the validity, consistency and accuracy of the findings (Merriam, 2009; Creswell, 2007; Mertens, 2005; Patton, 1990; Lincoln & Guba, 1985).

9. **External evaluator.** Means that the researcher invites other researchers to evaluate the logic used to interpret data from the study and the inferences reached. For example, if the data comes from video recordings, allowing the external evaluator to corroborate the interpretations of the researcher in light of the video. Likewise, allowing other researchers to evaluate the findings of the study tracking the creation of the categories from transcript information or observation memos. This technique helps to identify any logic error in the exercise of the researcher to interpret data, develop categories or make inferences. Another name for this technique is data audit. (Merriam, 2009; Creswell, 2007; Mertens, 2005). The objective of this technique is to increase the validity of findings allowing an external evaluator in helping the researcher to eliminate logic errors in the interpretation of their data.

b. Controversies with internal validity. Establishing the validity of the findings is the fundemental scientific basis of the researcher. The issue of internal validity in qualitative research sparks controversies for two reasons: There is no consensus on the criteria to be used to describe the validity of the findings of social phenomena. For example, in this book we talk about the internal validity while other authors use terms such as credibility (Guba & Lincolh, 1985) or the quality of the study (Patton, 1990). When these terms are examined, various criteria are identified to determine the value, consistency or accuracy of the data. Creswell (2007) recommends that qualitative researchers use those terms and criteria that best meet their study. When it comes to establishing the validity of the study, the questioning arises of how far should the educational researcher go to achieve: depth description vs. verification of the phenomenon.

1. **Depth Description of the phenomenon.** This position argues that the validity of a qualitative study rests on the ability of the researcher to describe the social phenomenon accurately. The argument is that social phenomena of education are mental constructs. These do not have physical properties that can be corroborated. Deep and accurate description that identifies and builds on the experiences of those who experience them, it is the only means of making concrete a social phenomenon that exists in the minds of people. The other challenge to internal validity is then writing a report that is clear, convincing and with evidence, where the reader can evaluate the findings and take their own stance on the validity of the research study. In this position, the concept of "verification" that permeates educational research entails a positivist tone and measurement which is not well accepted or considered practical when talking about social phenomena (Creswell, 2007; Van Manen, 1990; Strauss & Corbin, 1990; Lincoln & Guba, 1985; Glaser and Strauss, 1967).

2. **Verification of the phenomenon through its observable manifestation.** This position argues that the experiences of the participants should be checked, when possible. The argument is that if social phenomena of education are experienced it is because they transcend the mere appreciation of people (Pring, 2000; Miles & Huberman, 1995; Patton, 1990). The perception, when it becomes behavior, can be observed. The validity of the qualitative study is set when the social phenomena of education are translated into observable behaviors. There are qualitative studies in education where the goal of the researcher transcends description of these perspectives of participants to explain them in relations of cause and effect (perception-behavior). In this type of qualitative research it is necessary to identify the observable manifestations of the phenomenon to validate them. Qualitative researchers investigating social phenomena, and seeking to identify their observable manifestations, are considered operating from a pragmatic position (Patton, 1990) or realist (Miles & Huberman, 1995) research and qualitative evaluation. Others argue that this is a positivist stance because the social world is a mental construct (Denzin, 2009).

c. Closing Remarks on internal validity. Many qualitative researchers begin their studies in education from the premise of studying social phenomena of education. It is in the course of the study they discover observable manifestations of the phenomenon. Hence the recommendation of Van Manen (1990), qualitative research can not be limited to superficialities. The researcher has to find a way to ensure that they capture the phenomenon researched, be it social or physical.

d. External validity. Means the application of the study findings to other unstudied educational settings. For example, if the study is on successful management practices in an urban school, then how these findings can help other urban schools to be successful too. When qualitative researchers speak of external validity, it refers to a transfer of learning managed by a study to use it to improve educational practices in other schools. This is possible because qualitative studies focus on social phenomena in educational contexts defined as the classroom or the school. The effective transfer of learning from a qualitative study rest on two pillars: the phenomenon being studied and the context in which it manifests. An example of this would be the phenomenon of desertion. Desertion has two major components: the interpretation that the student has with their academic experience and home culture. If you understand the interpretation and school and family context in which it occurs, then that learning can help deal with desertion in schools that exhibit similar social, economic and educational conditions.

1. **Describe in detail the phenomenon being studied.** This is essential so that the reader can understand what phenomenon is being studied or what its manifestation is. For example, if the study addressed administrative practices, school dropouts or student retention.

2. **Specify and describe the context in which the phenomenon studied manifests.** The educational context is not only the physical environment of the school, also includes the social circumstances, cultures, politicies or economics surrounding the manifestation of the phenomenon in the educational institution. For example, the culture of the school and the community of origin of the students, income levels or social problems reflected in these. The educational context describes the physical and cultural environment of the institution or educational system where the phenomenon being studied is investigated. The detailed description of the context allows evaluation of the manifestation of the phenomenon being investigated and how it affects the same.

3. **Identify the learning of the study and how it can help improve education in those schools with similar conditions.** Explain what

was discovered about the phenomenon, how it arises, why, and what conditions of the school or education system favor or allowed it. In this way the reader can appreciate the learning of the study and use this to improve education in educational settings with similar conditions.

3

Mixed Methods Research

Omar A. Ponce and Nellie Pagan

Metropolitan University, PR, United States

3.1 Introduction

Mixed methods research begins with the process of thinking of the researcher and not the methods. A mixed mentality of research means the capacity of the researcher to understand and to use the plurality in the research. This mentality emerges in many disciplines of studies before a society increasingly more multicultural and complex (Greene, 2007).

Since the 1990s, an increasing number of researchers conduct studies where the quantitative and qualitative approaches of the research design are combined or are an integrated component.

This practice has been given the name mixed methods research. Several reasons seem to explain the rise of mixed methods research: The emergence of a global, multicultural and increasingly complex society. The complexity of social, economical, political and educational contemporary problems demand, increasingly, research of alternate approaches that allow entry to this new complexity. In addition, the recognition of the limitations inherent in its conceptualization, which as models presents approaches from quantitative and qualitative research to address the totality and complexity of human endeavor. Finally, within the social complexity that we live, emerges a more pluralistic and flexible view of research, where models of quantitative and qualitative research are recognized as complementary to each other to enter the social complexity in which we live. (Caruth, 2013; Ponce, 2011; Creswell, 2009; Greene, 2007).

Mixed methods research is recognized as a third model of research. It means that has its own characteristics and research designs (Caruth, 2013). Also, exists a recognition that the researcher of 21st century has to master quantitative and qualitative research models to enhance these in mixed methods research (Creswell & Plano Clark, 2011: Ponce, 2011). In this chapter the model of mixed methods research, their characteristics and basic research designs are explained.

3.2 Mixed Methods Research

A mixed methods study is research intentionally combining or integrating quantitative and qualitative approaches as components of the research. The use of these approaches can occur at different points in the research process. (Caruth, 2013; Creswell, 2011; Ponce, 2011; Teddlie & Tashakkori, 2009; Greene, 2007).

a. In the planning phase where the research plan is developed, it becomes clear what is investigated and how quantitative and qualitative approaches are used.
b. Combining or integrating research questions from quantitative and qualitative approaches to guide the researcher into the complexity of the problem studied.
c. Using quantitative measurement instruments with qualitative research techniques to generate quantitative and qualitative data for the research problem.
d. Combining or integrating quantitative and qualitative data in the analysis of study data.
e. Combining or integrating quantitative and qualitative data in the presentation of the study findings.

Four objectives are pursued in mixed methods research (Caruth, 2013; Creswell & Plano Clark, 2011; Ponce, 2011; Teddlie & Tashakkori, 2009; Berman, 2008; Greene, 2007):

a. Combining or integrating quantitative and qualitative methods toward the best possible approach to the research problem.
b. Generate quantitative and qualitative data toward a clear and deep understanding of the research problem being addressed
c. Generate quantitative and qualitative data from the same research problem that allows the researcher greater certainty in inferences, conclusions or statements which formulate its findings.
d. Make more robust research by using the strengths from one research model to offset methodological shortcomings from the other. This produces more reliable research.

3.3 Characteristics of Mixed Methods Studies

There are many forms of quantitative and qualitative research. Both the model of quantitative research and qualitative operate upon some assumptions about what research is and how it should develop. When the researcher combines or integrates quantitative and qualitative approaches in the design of mixed study, what it does is create a third research model that allows using these two in an articulated and harmonic manner. The first step to combine or integrate quantitative and qualitative approaches in the same study is to understand the assumptions, the foundations and characteristics of mixed studies, as a third research model (Caruth, 2013; Creswell & Plano Clark, 2011; Ponce, 2011; Campos, 2009; Morse Niehaus, 2009; Teddlie & Tashakkori, 2009; Greene, 2007; Mertens, 2005: Tashakkori & Teddlie, 1998). Let's review some of them:

a. **The nature of the problems of mixed research (premise).** Mixed methods research is used only when the complexity of the research problem cannot be addressed from the unique perspective of a quantitative or qualitative study. The argument is that contemporary society has evolved and has become more complex. The vast majority of the social, economic and contemporary political problems show that complexity. Let's use the example of environmental problems. Environmental pollution is the result of many factors; urban development, the development of various means of transportation, like the car, the use of water bodies as a recreational environment, and so on. Researching the problems of the environment, and venturing into its complexity, demands the use of multiple studies to address the complexity. Mixed methods studies are based on the belief that there are existing problems whose complexity cannot be fully researched when the combination or integration of quantitative and qualitative approaches are not undertaken as components of the study. Simply put, the complexity of the problem cannot be deciphered or fully understood from a single quantitative or qualitative approach. Mixed studies address research problems in which clear objective and subjective aspects are manifested that require the use of quantitative and qualitative approaches. For example, the temperature in a cinema. Subjective elements are aspects of the problem that can be understood only by the perceptions and experiences of those who live them. For example, if the temperature of the theater is considered pleasant or unpleasant. Mixed methods research is used only when we address research problems which have objective and subjective elements in its manifestation.

b. **The research question (foundations).** Mixed studies emphasize the research question of the study being the focus of all methodological decision. The research question guides the study and determines which components of quantitative and qualitative models are used. In other words, what determines the combination or integration of quantitative and qualitative approaches are the research questions of the study. The relationship between the research questions and the quantitative and qualitative approaches should be seen clearly when designing the study. This is important to establish the relevance and alignment of quantitative and qualitative approaches which are selected to study the research problem.

c. **The process (methodology).** The process of a mixed methods study is to integrate or intentionally combine quantitative and qualitative approaches as components of the study. The aim is to explore the complexity of the research problem to measure their objective aspects and to understand/describe their subjective elements as directly and accurately as possible towards its manifestation or expression. The combination of approaches occurs in two ways; prior to the study or in the planning stage as in quantitative studies, or in the development of the study where the researcher recognizes the need to depart from the original design of research to achieve their research goals, as in qualitative studies.

d. **The behavior of the researcher (philosophy).** The action of the researcher is pragmatic, meaning the product is more important to study the process. Any decision on how to combine or integrate quantitative and qualitative approaches, once the study is in place, is based on how these provide an insight to the complexity of the problem and answer the research questions of the study to achieve the research objectives. The more a combination or integration of quantitative and qualitative approaches can zoom in and capture the essence of the problem, the greater the relevance and effectiveness of the design. When this occurs, it can be argued that the researcher's decisions were correct.

e. **The study product.** The product of a mixed methods study is quantitative and qualitative data upon the problem studied. The collection of quantitative and qualitative data provides more complete information, descriptive or broader, from the research problem and this allows the researcher to make more informed decisions about how to solve the same.

3.4 Models of Mixed Methods Research

There is no universally accepted definition of mixed methods research. In the literature the following two models mixed methods research are identified (Ponce, 2011; Creswell, 2009):

Model 1

1. Is the first model of mixed methods research identified in the literature prior to the 1990's.
2. Quantitative and qualitative approaches in the same research were used, but not connected, integrated or combined.
3. Combining or integrating the data is done at the end of the study to answer the research questions.

Model 2

1. Is the model of mixed methods research emerging in the 1990's, and defines its contemporary practice.
2. Combining or integrating research approaches intentionally to produce a more robust study that one of mono-methodological approach.
3. Integrating approaches occurs in the philosophical positioning of the study, methodology and data analysis.

3.5 The Basic Structure of the Mixed Methods Study

Two basic structures or ways of combining or integrating quantitative and qualitative approaches as part of the design of a mixed methods study are recognized. These structures are explained below (Caruth, 2013; Creswell & Plano Clark, 2011; Ponce, 2011; Teddlie & Tashakkori, 2009):

a. **Research in sequential phases (*sequential phases design*).** Signifies that the researcher begins his study with a research approach (phase I) and uses findings to design a second phase (Phase II), but using another research approach. For example, the study begins with a qualitative phase and uses findings to design the quantitative phase. The fundamentals of studies with sequential phases are to use a research approach to study deeply the research problem and then use the findings of the first phase and design the second phase. The two possible combinations under the structure of sequential phases are presented in Figure 3.1:

b. **Research in parallel phases (*convergent parallel design*).** Means that the researcher uses quantitative and qualitative approaches simultaneously in the development of their study. Generally, parallel phase

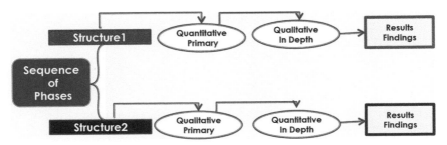

Figure 3.1   Sequential phases mixed studies structure.

studies consist of studying the problem in an integrated manner from the quantitative and qualitative approaches. Figure 3.2 illustrates the structure of parallel phases of a mixed methods study.

3.6 Mixed Methods Research Designs

The design means the research plan that will guide the researcher in conducting the study. Mixed research designs are accepted ways of how integrated quantitative and qualitative approaches can be combined in mixed methods study. Below are presented seven mixed methods research designs that illustrate the structures of sequential phases and of parallel phases (Creswell & Plano Clark, 2011; Ponce, 2011; Teddlie & Tashakkori, 2009; Greene, 2007):

a. **Exploratory design using sequential phases (quantitative - qualitative).** The objective of this design is the exploration of the research problem. Exploration is used when very little is known about the research problem. The less information about the problem, the greater the relevance of this design to begin learning about it. This design first uses a qualitative research approach to explore the experience of participants with the phenomenon under study, their culture or values

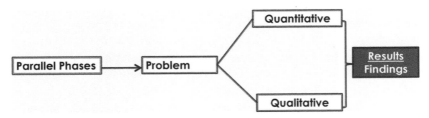

Figure 3.2   Parallel phases mixed design structure.

of the group, or the structure of the institution being studied. With the findings of phase I (qualitative), the researcher designs a quantitative study (phase II) to define or measure the findings of the qualitative phase (phase I) in a sample of the universe under study. For example, say that a car company wants to redesign your model sedan in the face of reduced sales. As they do not know the needs and interests of their customers, the study will begin with a qualitative approach using focused interviews. Identify buyers of that sedan model of the San Juan dealer, start the interview process and generate a list of the aspects both positive and negative and should consider the recommendations for a sedan car more responsive to the interests and needs of customers who bought it. With these findings, the researcher designs a questionnaire (quantitative phase II) to be administered to a sample of buyers of the same model at other dealerships of the company. Once administered the questionnaires and the data collected can specify the order of preferences, strengths and needs of customers to be incorporated into the next model line sedan cars offered for sale. In this study, the qualitative phase reveals the needs of customers and the quantitative phase facilitates the understanding of these needs in a large sample of the population (Figure 3.3).

b. **Explanatory design using sequential phases (quantitative - qualitative).** The purpose of this design is to study or describe the research problem in depth. To achieve this, it first uses a quantitative study to measure the attributes or properties of the problem (phase I) and then to a qualitative study (phase II) to deepen the findings of Phase I. For example, say that the Residents Association of an apartment complex decides to study the levels of resident satisfaction with recreational areas. A survey with questionnaires to residents is performed (phase I). This survey asks how satisfied residents are with recreational areas,

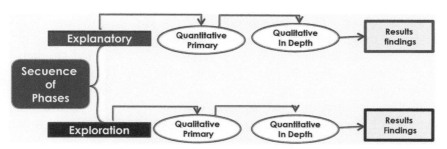

Figure 3.3 Mixed design basics.

using a scale of 1 to 4. In this survey, 1 means very dissatisfied and 4 very satisfied (quantitative phase). After the structured quantitative survey the study continues with a qualitative interview, this time trying to understand the reasons for the initial response (qualitative phase). Each resident is asked to explain or qualify his answer. This allows generation of a list of strengths and weaknesses as perceived by residents of the recreational areas of the apartment complex. With this data, a plan is generated to meet the same. In this example, the quantitative phase measures the level of satisfaction of residents with the recreational areas and the qualitative phase allows us to understand the reasons (Figure 3.3).

c. **Convergence design using parallel phases.** The objective of this design is to study the research problem in its entirety and dimension. The quantitative approach is used to measure the properties and objective aspects of the problem. The qualitative approach is used to understand and describe the subjective aspects. It is known as convergence because each design approach is used to study different aspects of the problem. The quantitative approach measures the objective aspects of the problem and the qualitative phase enters the subjective aspects of the problem or the experiences of the participants. Convergence occurs because it is the researcher who integrates quantitative and qualitative data to explain the problem studied. For example, let's say the X Hospital hires a consultant to assess their pediatric services. As a quantitative approach, the consultant designs a checklist for pediatric services regarding to the commitments established by the hospital toward patients; no wait time more than 30 minutes to be served, friendly doctors and nurses, make the best treatment accessible to patients, based on income and people's health plan. With this checklist, the consultant observes the performance of the pediatric ward, and how they care for patients and evaluates services according to established commitments. To understand the experience of parents and children with pediatric services, interview those who agree to be interviewed. In this conversation, the researcher tries to understand the perception of parents and children with the service they received. With the quantitative assessment component measures the degree of concordance between the services offered with the commitments in the hospital's mission. With the qualitative component, understands the experience of parents and children with the service received. The convergence of quantitative and qualitative data allows the consultant to explain how responsive and consistent pediatric services are offered by the hospital with the promises it makes to their patients, according to the experience and lives of the patients.

d. **Triangulation design using parallel phases.** The objective of this design is to use quantitative and qualitative approaches to study in depth the same aspects of the research problem. To achieve this, the researcher carefully plans the entire process of research to address these aspects of the problem from quantitative and qualitative perspectives. This is achieved if the measuring instruments and research strategies are aligned and complementary to collect quantitative and qualitative data of the problem. Thus, the data analysis focuses on these aspects to obtain quantitative and qualitative data to triangulate or consider the same aspects of the problem. We return to the example of the consultant evaluating pediatric services. To make the study of Figure 3.3, a convergence study one of triangulation, the consultant must define pediatric services, build a checklist, align this to the promises (mission and vision) established by the hospital to their patients to measure the performance and effectiveness of services (quantitative phase). Also he needs to know the experience of parents and children, with each of these services received (qualitative approach). In conducting the study, he uses quantitative and qualitative approaches to examine the same aspects of the research problem. Thus, the researcher is able to penetrate and explain the problem in depth from quantitative and qualitative perspectives (Figure 3.4).

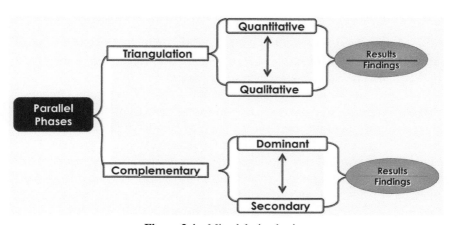

Figure 3.4 Mixed design basics.

e. Complementary design using parallel phases (*embedded designs*). The objective of this design is to use one of the research approaches to counter the deficiencies of the other. In this design, a research approach is used in a primary role because it is the dominant or principal method of study. Let's say we use an experiment as the main research method to test the effectiveness of a technique for relaxation and stress management. The strength of the experimental design in this case is that it tests, with people, the relaxation technique. Typically, the stress level of the subject before the test is measured, the technique is applied and the result is measured to determine whether the treatment effect appeared or not. The test technique with people is the greatest strength of the experiment. However, its shortcoming is that it does not provide an explanation of the process or how the technique works. This occurs because the experiment assumes that if a change occurred in the levels of relaxation of the subjects, as measured in the pre and post-test, it was due to technique. In this example, if a qualitative research approach is used where you can understand the experience of those with the technique is incorporated, then you might have a clearer idea of how it works. In this case, the experiment is the primary method of research and the qualitative approach is the complementary method because it is used to compensate for the methodological deficiencies of the experiment (Figure 3.4).

f. Multilevel design (*multiphase design*). Multilevel designs are studies where the researcher needs to venture to different levels of analysis, study and research because the problem has several dimensions, manifestations or ramifications. Therefore requires different research approaches and different groups or samples to enter this complexity and to decrypt it. Let's say that the management of a private school decides to adopt a new curriculum where all classes will be conducted in English and Spanish alternately. One day the class will be in Spanish and the next in English. The school administration argues that this new approach will help create bilingual graduates better prepared for many changes that are occurring in the workplace. To assess the extent of this decision prior to implementation, the measure should be understood from different perspectives of what it will entail for faculty, students and parents. For example, how would this influence the daily preparation of teachers to teach a course bilingually. The same would apply to students. What monetary cost would this decision have for parents, perhaps, to purchase materials and equipment for their children in English and Spanish. In this example, the list of questions (complexity of the problem) can be

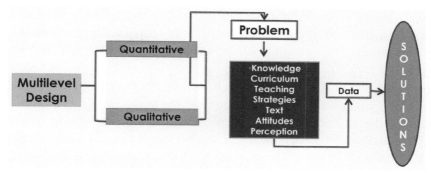

Figure 3.5 Multilevel design of a problem.

much larger, if approached from the various population groups that would be impacted by the decision (parents, teachers, students). To study a problem of this nature, it is necessary to use a multilevel design or a study that uses various quantitative and qualitative approaches using different population groups or samples, as part of the research design (Figure 3.5).

g. **Emergent designs (*transformative design*).** It is common in mixed studies to deviate from the research design for the following reasons that occur when combined quantitative and qualitative approaches in the same study; the researcher encounters quantitative and qualitative data that contradict, the researcher identifies new perspectives on the problem that had not been included in the initial study design, however they merit investigation, or discovers methodological errors in the study. When these situations occur, the researcher has two options; concludes his study and accept this as limitations of research or modify the design to respond to them. When these modifications are made to the joint studies to answer those findings that emerge from the research process and that merit response it is known as an emergent design. Therefore, the researcher must explain this in the final research report. In other words, the study began with a design that evolved to another in the process of conducting research.

3.7 Data Collection in Mixed Studies

Data collection constitutes the phase of developing mixed methods research. The fundamental principle is to collect data **respecting the rules of each research model in the design and developing of a mixed methods study.**

What is quantitative is quantitative and what is qualitative is qualitative. Keeping each research approach aligned within their paradigm or model strengthens the rigor of each approach and the validity of the mixed study.

a. **Define the research problem.** The research problem consists of situations, phenomena, processes or persons who are the focus of study. In mixed studies, research problems have the tendency to be complex because they include objective and subjective elements to be addressed with a combination of approaches. There are two styles when presenting the research problem:

- **Write the problem by way of composite question where the objective and subjective aspects are highlighted.** For example; Which candidate for governor people prefer and why?
- **Write the problem by way of simple question and leave the objective and subjective aspects to the research questions.** For example; What radio station do those aged 20 to 30 prefer?

b. **Write the research questions.** Research questions decompose the problem into manageable units to be studied. In mixed methods studies quantitative and qualitative questions are used. A common practice in mixed methods studies is always designing questions beginning with what, how, when and where. This is accepted because it is easier to answer questions when contrasting with the survey data. For example; What is the social issue that most worries Puerto Ricans? Two styles dominate in drafting research questions in mixed studies:

- **Write research questions for each research approach.** In other words, write four to five questions for the qualitative component and the same amount for the quantitative component. In this format, each research approach answers its research questions. Those who favor this format argue that it provides much more specificity to the research because each component has its own questions.
- **Write research questions to guide the entire study.** In this format, the full study aims to answer these research questions. In other words, the quantitative and qualitative components of the study are designed to generate quantitative and qualitative data to answer the research questions.

c. **Select the research design.** The principle key in selecting the design is to understand the quantitative and qualitative research to use them appropriately in a mixed methods study. It is very difficult to conduct a mixed methods study without understanding the models of quantitative and qualitative research.

- **The problem and research questions have to connect with the mixed research design.** For example, studies of exploration or explanation. Studies of this nature can be carried out with sequential phase studies. If the research questions put greater emphasis on one component of research, it can probably be done as a complementary design using parallel phases. If the research questions put equal weight on quantitative and qualitative models, then it could be answered either with a convergence or triangulation design using parallel phases. The research questions are fundamental in determining what and how the approaches of quantitative and qualitative research were used.

- **Specify what to combine, integrate or complement and why.** The argument for using mixed methods is to enter into complex problems. To accomplish this, you can combine quantitative and qualitative models to examine the objective and subjective aspects of the problem. It is possible to combine the research questions with instruments and data collection techniques to generate quantitative and qualitative data that allow a deeper description of the research topic from a mono-methodological or a single quantitative or qualitative perspective study. There must always be a logic that allows explaining what and why the research questions were combined. This logic must be based on the relationship between the research aims and how this allows the achievement and success of the study. Nothing can be a whim of the researcher at the time of combining or integrating quantitative and qualitative approaches.

d. **Write the study title.** Titles should reflect three components; the research topic, the study population and research design. An example of a title could be: Factors that influence people to visit a mall: Exploratory mixed study in sequential phases (Creswell & Plano Clark, 2011).

e. Select the sample. In mixed research methods two types of sampling dominate (Ponce, 2011):

- **Primary sampling (adhere to the established).** Consists of selecting the sample according to the parameters of the respective models of quantitative and qualitative research. The researcher selects samples and does not deviate from these.

- **Alternate sampling (deviating from the established).** In mixed studies, three phenomena occur which force the researcher to deviate from the research plan; encounter quantitative and qualitative data that contradict, discover methodological gaps in the study due to the nature of combining approaches, as would be discovered in the interview process (in the qualitative phase) that the questionnaire used (in the quantitative phase) does not address the whole issue or new issues emerge that it is necessary to study. When these situations occur, the researcher has two options, accept these as limitations of the study or deviates from the original research plan to compensate for them. To address and resolve these situations the researcher must employ the strategy known as alternate sampling; by selecting additional samples. Alternate sampling in mixed studies are criterion samples, as in qualitative studies, or the selection of samples that allow answering the research questions of the study.

f. **Develop tools and research techniques.** As the sampling, the development of tools and techniques for data collection must adhere to the criteria established by the quantitative and qualitative models. An important element in this task is to ensure that the tools and techniques of data collection are aligned to the research objectives; generate the quantitative and qualitative data to answer the research questions, generate quantitative and qualitative data to understand clearly and deeply the research problem, produce quantitative and qualitative data of the same phenomenon under research.

g. **Address individual authorities to conduct the study.** In conducting the study, follow channels or procedures and comply with the provisions of the agency or institution where the study is to be conducted.

3.8 Analysis of Mixed Data

Analyzing data is to extract meaning, implicit or explicit, of the information collected in the study. Analyzing data is a three step process; encode and describe the information to understand the messages that may be there, analyze and interpret information to make it clean data and communicate findings and identify the most effective way to convey the findings. In mixed studies three types of data analysis are used; analysis of quantitative data, qualitative data analysis and analysis of mixed data. The analysis of mixed data consists of organizing and combining quantitative and qualitative data to achieve one or more of the following objectives related to the research topic:

   a. Triangulation of data. Is demonstrating how quantitative and qualitative data collected in the study are validated between each other. Triangulation means that the quantitative and qualitative data match, point in the same direction or converge on aspects of the research problem. For example, in a study the quantitative and qualitative data show that the participants enjoyed the educational conference given to them. The average evaluation was 3.5 on a 4.0 scale. Comments on this conference were to be repeated again and invite the family members. Triangulation of data is possible if the measuring instruments and techniques of qualitative data collection were designed to collect quantitative and qualitative data from the same aspect of the research problem. When the researcher has data to triangulate, they increase the validity of the study and facilitate inferences and conclusions that can be stated about the findings.

   b. Complementing data. It means using quantitative and qualitative data to complement when presenting the findings. Complementary signifies that data supports each other. For example, 90% of individuals classified the film as excellent (quantitative data). The reason for this was that they found it fun, educational and suitable for the whole family (qualitative data). In this case, the quantitative data sets the scope of the measure and the qualitative data deepens it or one data set complements the other.

   c. Deepening in data. Signifies using quantitative and qualitative data to bring the argument to a point of no refutation. In this analysis, the amount of quantitative and qualitative data provides an overview of the research problem. While in "triangulation", quantitative and qualitative data point in the same direction, in "complementing" a set of data supports the other; in "deepening" the quantitative and qualitative data provide a comprehensive and clear view of the research problem. For example, in a study we found that 90% of parents and 95% of students did not endorse the new math curriculum (quantitative data) because they consider it too complex and impractical for developments in teaching this subject (qualitative data). Comparing this curriculum with the American standards for the teaching of mathematics, practices and teaching strategies are clearly identified that contradict the established professional standards for a modernized teaching of mathematics (qualitative data). As shown in this example, the quantitative and qualitative data are used to bring the argument to a point where there is no doubt regarding the findings of the study or to a point of no rebuttal because a comprehensive picture of the topic is provided.

3.9 Validity in Mixed Studies

In research, terms like internal and external validity are commonly used to describe the investigative rigor of a study. The term **internal validity** is used to describe how much correspondence exists between the data collected and the research problem. **External validity** refers to whether the study data can be used beyond the context of the study or applied to other samples that were not studied. In mixed studies the validity criteria from qualitative and quantitative models are used to meet the investigative thoroughness of the respective models. However, as mentioned above, the aim of combining or integrating quantitative and qualitative approaches is to venture into complex problems where there are clear objective and subjective aspects to generate quantitative and qualitative data to more or better approach to the research problem. In mixed studies the term of **inference validity** is used to describe the effectiveness of the researcher to approach and capture the complexity of the research problem using quantitative and qualitative approaches. Inference validity signifies that the quantitative and qualitative data describe, explain or accurately capture the research problem and its complexity. When this occurs, the researcher can argue that it was effective in combining or integrating qualitative and quantitative approaches and is therefore in a better position to make valid inferences or interpretations of the research problem for the richness of its quantitative and qualitative data collected. Below are several recommendations to establish the validity of inference from a mixed methods study:

a. **Compliance with the validity criteria established in each research model.** Always seek that the quantitative and qualitative approaches from your mixed methods study meet the criteria for internal or external validity of their respective models. Ensure that the internal and external validity of each research approach contributes to the validity of inference of the mixed study.

b. **Establish the conceptual validity of the research problem.** This signifies that the research problem is really a problem for mixed methods research with objective and subjective elements. Is difficult to measure or describe if an attribute does not exist in the research problem. At the time of writing the research report, the research problem becomes clear, its complexity and its objective and subjective aspects.

c. **Establish the methodological validity of the mixed study.** This means to clearly establish the alignment of the selected mixed design with the research questions and the objective of the study. It should clearly explain the relationship between the objective of the study, the research questions and the mixed design. If the mixed study deviates from planned, is necessary to clearly establish the researcher's logic or rational use with the emergent design. The aim is that the reader of the study may assess the researcher's procedures and determine the validity of the research process or emergent design.

d. **Establish the validity of the research product.** The validity of the product is evaluated on the relevance of the data analysis and the correspondence between the data collected and interpretations made by the researcher of this information. The relevance of the analysis of the data signifies that the techniques used to analyze the study data correspond to the information gathered by helping to interpret accurately. For example, in quantitative research, analysis of standard deviation is a truer measure of dispersion than range analysis. In qualitative research, validation with the participants of the significance of the categories that emerge from their interviews results in a measure of greater certainty than what may be provided by an external evaluator/consultant. The respondent is in a better position to clarify what he meant about the theme instead of the opinions of an external evaluator about the category used by the researcher to describe the intent of the interviewee. The correspondence between the survey data and interpretations made by the researcher signifies the researcher's ability to cement each interpretation with the study findings. Although this is an exercise in logic, the researcher clearly establishes the link between the data and their interpretation. In mixed studies, this link is expected to be a compelling one because the researcher collected quantitative and qualitative data for the research problem. Therefore, their margin for error in interpreting should be less because it has two sets of data to formulate their interpretations and recommendations.

3.10 Writing the Mixed Methods Research Report

There is no universally accepted way of how to write the mixed methods research report. Below are presented several recommendations on mixed methods research reports, especially in theses and dissertations:

a. The content of mixed methods research reports follow the same linear deployment of quantitative and qualitative thesis or dissertations; Statement of the problem (Chap. I), literature review (Chap. II), method (Chap. III), findings (Chap. IV) and discussion (Chap. V). The challenge in presenting the contents of a mixed methods report is to let the reader see clearly, and in an orderly manner, the type of study that was conducted; sequential phases or parallel phases. The reader must understand quickly how the qualitative and quantitative approaches were combined or integrated in the mixed design research. The objective is to convey the feeling that two studies are presented in a single report. This challenge is evident in the writing of Chapters I, III, IV and V.

b. In presenting the research problem in Chapter I, it must clearly establish the complexity of the problem and justification for a mixed methods study. The problem may be complex, however for a mixed methods study the objective and subjective criteria must be categorically established. The other challenge in presenting the research problem lies in the way the research questions are presented to coordinate the study. The challenge here is whether research questions are presented to guide the entire study, or make mixed research questions to guide the quantitative and qualitative phases of the study. The selected style to present the questions facilitates in articulating how these are connected with the mixed research design and guide the study. The clarity and precision of Chapter I facilitates the development of the remaining chapters of the report. Another consideration when presenting the research problem is whether the study deviated from the initial research design. The explanation of the emerging design is done in a section entitled "methodological considerations." This section may explain details such as the selection of emerging samples, changing measuring instruments or other methodological decisions that led to deviate from the initial research plan. This section should not be confused with the sections of boundaries and limitations of quantitative studies.

c. In Chapter III the challenge is to present, in a consistent manner, the combination or integration of qualitative and quantitative approaches as a mixed research design. Our recommendation is to present the chapter corresponding to the type of study presented; sequential phases or parallel phases so that the reader can understand the development of the line of study. For example, if the study is sequential phases, then fully explain phase I and later phase II revealing to the reader how each phase is connected with the other and thus constitutes a combined study. Avoid presenting each approach as if it were a separate chapter of another.

d. Organize and present the findings of the study in a way that allows answering the research questions. The clarity of the presentation of the findings is greatly facilitated by the selected strategy to communicate information. For example, if tables or graphs to summarize data or integration are used. Tables are an excellent strategy in mixed studies to summarize and integrate quantitative and qualitative data visually or on different aspects of the same problem. The goal in presenting data should be to communicate these clearly and accurately where quantitative and qualitative data facilitate answering the research questions.

e. The wording of the report should contain language that handles each research model. In other words, the presentation of the quantitative phase must conform to the technical language of the quantitative model and the qualitative phase model has to conform to the technical language of qualitative model. This is critical so the mixed method researcher demonstrates understanding, dominance and respect of the respective models rules and practices.

3.11 Closing Remarks

In this chapter the fundamentals of mixed studies are presented. They are called mixed research studies where quantitative and qualitative approaches are intentionally integrated or combined as components of the design. Although mixed research designs are recognized, the essence of integrated or combined research approaches lies in having a joint research mentality. That mixed mentality emerged greatly to understand the potential of quantitative models and qualitative research and provide opportunities to integrate both.

4

Methods in Educational Technology Research

José Gómez Galán

Metropolitan University, PR, United States,
and University of Extremadura, Spain

4.1 Introduction

Contextualizing educational technology in the field of scientific research we are at a moment of maximum interest in the field of Educational Technology. The digital revolution has led to an extraordinary development calls Technologies Information and Communication Technologies (ICT) with all didactic and pedagogical implications of this complex phenomenon have in the world of education. Parallel to this, the Educational Technology has evolved dramatically from its origins to the present, especially in a world like today, dominated by the digitization of information and communication processes.

That is why we consider essential, and will be the main objective of this work, analyzing the most prominent paradigms of scientific research in the field of Educational Technology, have emerged in recent decades as responses need to demanding educational contexts in a gradually more technical and mediated, especially focusing on the newest and innovative proposals that today are aimed at addressing the challenges posed by the ICT revolution and its influence and impact on society formation processes.

The transformations of educational research within the Educational Technology is present in both theoretical studies and the field work, and forced the development of new models that allow the needs generated by new teaching and learning environments, each increasingly widespread in the international scientific arena.

In this sense, regardless of classifications and taxonomies of classical scientific research in educational technology, in our case we will establish two major dimensions that locate the main models and paradigms: on the one hand we call intraeducacional educational technology, centered internal problems of education-primarily didactic, organizational and methodological-, and other educational technology supraeducacional in relation to other social systems, or in other words, what the new information society will-supposed-to our world and coping from an educational perspective. Within these two large present different paradigms and educational models.

Believe necessary and so take out in this study, analyze and compare different methodologies, techniques and scientific strategies of Educational Technology in the field of social sciences, humanities, experimental sciences, affecting both its historical development and its situation right now, in the digital environment- society, so that the scientific and educational technologist can have a complete overview of the main existing research paradigms in order to guide their scientific work, depending on your needs, in this important object of knowledge. Also place them in a historical perspective allows us to check the validity and relevance of many of them, or the need to adapt to new socio-educational realities. It is one of the fundamental goals that we pursue.

4.2 Challenges of ICT Research: Response to New Socio-Educational Needs

The emergence of ICT in education has implied the existence of variables and unknowns so far, which make research models might be called traditional insufficient, of course, to access an empirical development to achieve objectives having a direct or indirect relationship with the influence or presence of these technologies in the educational world.

From a traditional perspective, the research developed within science education had focused on studying, analyzing, controlling and/or apply different educational phenomena, for many decades, were quite homogeneous. However, the explosion of ICT during the last thirty years has meant that the dimensions listed above, both supraeducacional, relationship with other education systems (social, political, economic, cultural) systems like intraeducacional, should be seriously reviewed from the perspective of educational research.

In that context, of course, different questions arise to which we must respond: what models are, therefore, suitable for carrying out a process of research into new forms and needs of education systems in the presence ICT is increasing? How can we know the best strategies for integrating new technologies in education, according to urgent training needs of a society that increasingly depends on them? How do you get the implementation process is adequate, it is established in the form, manner and correct for it to be assimilated by some as tight and inflexible systems now, but apparently not submitted in this way, as educational? The answer is, of course, research frameworks have adapted to the new reality, which allow for an efficient way to obtain information, the analysis and evaluation of it is effective and accurate, and that the results allow implementation of correct and productive decisions.

However, the design of new models of educational research involves great risk and danger. A leap could mean more harm than current models continue to apply, even if they are displayed as many times as inadequate and obsolete. The strategy to follow, we must be adaptive. Using existing models and modifies them according to the new educational framework that is generating ICT. From this structure the effective models that we apply today should arise.

4.3 Classifications and Taxonomies of Classical Research in Educational Technology

To promote the integration of new technologies in an appropriate manner is necessary to use the experience gained in the process of a similar nature. We consider it appropriate strategy in the field of educational research; we must establish a new global ranking methodology. In principle we can establish that educational studies on new technologies have traditionally been referred both from a quantitative perspective, qualitative or, especially in recent years, mixed. In this sense, and in relation to ICT, there should be no way a priori no reason to set preferences. It depends on the nature of the investigation, if it is a desk or, conversely, if it is applied research. However, some authors (Gay, Mills and Airasian, 2011) have established a methodological categorization according to whether the research involved qualitative or quantitative paradigm. In this connection, we could establish that a possible initial classification should be adapted to the current needs in the object of our interest, and it could be [a] qualitative approach would seek a comprehensive understanding of educational phenomena studied through an inductive process. The methodologies of educational research, and would be today, for example, historical-educational research or qualitative research specifically; and [b] quantitative approach, which would involve the development of numerical data to explain the phenomena studied, essentially from a statistical analysis, a deductive methodological process. In this category we could include, for example, descriptive research, correlational research, comparative research, or experimental research.

Only course would be an example of the vast body of research that we provide an overview of the research possibilities of the problem in the scientific context of the social sciences. For an approach to the methodologies that are used today in the broad field of knowledge can be very useful inputs from Cohen, Manion, and Morrison (2000), Anderson and Arsenault (2001), Bless and Higson-Smith, C. (2004), Bryman (2004) and Punch (2005), Johnson and Christensen (2007), Houser (2008) and Creswell (2003 and 2012). And specifically for qualitative methodologies, which are undergoing further evolution are interesting jobs Maykut and Morehouse (2003), Charmaz (2006), Denzin and Lincoln (2008), Creswell (2012), Bazeley (2013), Saldana (2013) and Ponce (2014).

In recent years, due to the wide variety of objects of investigation, especially in the field of applied research and taking already into account the presence of ICT in education have been developed methodological preferences that would be added to classification we have offered to Gay, Mills and Airasian (2011), partly tangentially, are integrated into three major areas: *evaluation research*, focused on improving decision-making of the object studied; research and development (R & D, *research and development* of English), for the effective treatment of educational-products used or capable of being-in school contexts; and especially *action research*, which attempts to solve practical problems of the teaching- learning process using the scientific method. Especially in these areas there are significant studies on the methodological development (McNiff and Whitehead, 2002; Kincheloe, 2003; Flick, 2004; Gorard, 2001; Gorard and Taylor, 2004; Sagor, 2005, Thomas and Pring, 2003, etc.).

Found naturally consider other methodologies suitable for the processes of educational ICT research. However we note that the choice by the scientific education for whatever response should be, of course, why. I mean, what are the reasons that lead me to undertake this research, why I do it. We are talking about the function or purpose of it, as there are other plots within the research closely linked, as when (time) or what (object), among others. Just answering the main question would be possible to establish a general framework of research methodology should be performed, a successful and productive manner, the research work. However, we believe that to do it in an effective way we have to focus in an independent manner in the two dimensions that note, the intraeducacional and supraeducacional, which underpins our proposed classification staff, and ideally they consider them as a single and indivisible whole as truth is to systematize the research properly your particular analysis is needed.

4.4 Intraeducacional Educational Technology (IET)

4.4.1 General Features

As we have presented, the first major dimension is what we call *Intraeducacional Educational Technology (IET)*, i.e., the study of the integration of ICT in the educational process from a methodological and didactic perspective. They are, in short, they referred to as teaching resources.

The presence of new technologies in school contexts has led to the presence of a new element, without being radically revolutionary as suggested in some studies (Negroponte, 1995), it must be analyzed and researched in the new settings in which located. In the case of educational research different approaches exist that require individual research methodologies (curricular, educational psychology, educational, methodological, evaluative, organizational, etc.). Dependent on their areas of belonging to science education, but understood in its set as those that are directed to meet and resolve the problems arising from the direct presence of the technological tools in schools, both in the classroom and in the management and administration units. From this dimension, we consider intraeducacional would be necessary to establish the context and development processes in the same research can be carried out.

Naturally all these features involves systematizing only a guidance to facilitate the understanding of a complex field, thus approaching a traditionally more widespread and concepts used in epistemology and in particular educational terminology. Synthetically we could say that the main role so far of ICT in school- though not yet be fully extended, is the *teaching resources*, i.e., functions or teaching assistants learners. Furthermore, their presence as *instructional systems-* for example, to *process e-learning, m-learning, MOOC,* etc-., Or as support for administrative tasks involves independent study in the first place because it is a specialized field clearly, and in very specific educational settings, or by fitting in, in the second, with parallel processes to essentially pedagogical research, and closer to the management and economics. The TEI research, therefore, must consider the importance of ICT to facilitate and optimize their presence in the EA processes, self-learning, motivation, application of innovative teaching methodologies, etc.

4.4.2 ICT as Learning Resources for Learning in Techno-Media Convergence

ICT dominate all processes of information and communication in the world, and this is due to the digitization of information. All media and old analogue media have been replaced in the digital paradigm, by one means which converge all communication systems created by humans since their origins. This phenomenon, which we have called techno-media convergence (Gómez Galán 2003), no longer provides a separation between what was traditionally informatics and telematics, or between traditional media (film, radio, television, press, etc. .) and new (information systems and personal communication), but presuppose a permanent connection to information networks with the existence of a single digital medium in which they are all merged. Tablets, smartphones, etc., Are signs that the process of techno-media convergence, as announced over a decade ago (Gómez Galán 1999) is not only unstoppable and will lead us to a future of full connectivity, but that are already part of our present, are already the social reality in which we live. Today, ICTs are tools aimed at professional contexts: they are very powerful mass media.

Taken this into account, it goes unnoticed in many studies on new technologies that have completely independent of other media, the really characteristic of these is now in use in fundamental ways to promote or achieve different learning content - be they conceptual, procedural, or attitudinal, to help students in acquiring different skills to ultimately achieve different educational objectives. I.e., they are used primarily to facilitate learning, not to be confused with education-student. And this is the basic object of research TEI.

Learning is therefore based on these studies. In this sense the historical evolution of those investigations have relied on various learning theories, with the challenge of a global search theory. In particular the theories that have influenced the teaching of ICT evolve in parallel to the development of the same. Today it is easy to see too many perspectives: its presence in the teaching-learning process, its elements and relationships, how to present the information, its relevance to existing methodologies, or the creation of other specific-use, quality of materials, the convenience of when, how, and why use his presence in the classroom, etc.

Along with communication theories and systems theories are those influenced and are influencing the use of ICT as teaching resources. Thus, it is not possible to speak of a unified theory, agreed by the scientific community-which is not surprising due to the high number of variables that occur in the field of science education, which is not true in or experimental sciences, where an infinitely greater number of theories exist in the true scientific sense.

Educational research in ICT in the field intraeducacional, still involves educational technologists in trying to adapt the use of the same in the classroom teaching-resources-such as learning theories, communication and systems. In the early forties of the twentieth century in the context of communication theory, the work of Shannon and Weaver (1949) led many researchers consider to employ a means of communication could be as effective as direct teaching teacher-student.

But do not take into account variables such as teaching strategies employed, motivating teacher and student, cultural diversity, the time available for teaching, etc. (Hackbarth, 1999), which bound, without a doubt, learning, and were evidently present in educational contexts. Therefore, and from this perspective, the results were not as expected. In the early fifties, Schramm (1954 and 1957) sought a solution by modifying the basic model of the communication process and adapting to education, where highlighted two key elements: the form of messages and the role that experience means communication.

A wide field was thus opened within the emerging educational technology in which the ability of teachers and students, experience, or coding of messages, including derivatives thereof, and that conditioned the processes of teaching and learning, not are more complex communicative processes-they occupied a preferred position. Learning theories therefore became critical to understand the proper role of technology and media in teaching. As Mateos (1991) asserts, soon realized that the analogy-mind communication channel was insufficient, which led them to seek other paths. And in this sense we refer mainly to the two major paradigms generated during the twentieth century: behaviorism and constructivism.

Therefore, virtually all developed during this lengthy period of time can be applied today, because ultimately, and more embedded in the process of techno- media convergence in which we find ourselves, ICTs are only powerful media, not only of information as too often is incised.

4.4.3 Major Paradigms TEI: Evaluative Studies, Comparative Studies and IM / ATI Studies

Given the above we are at the moment to propose methodological models that identify, at present, with greater presence in educational research in ICT. The study of media as teaching resources is the perspective that has dominated research in education the same throughout the last century. We can group them into three sections: the Evaluative Studies, Studies and Comparative Studies Intra-Medium (we call IM) and research Aptitude Treatment Interaction (ATI).

The latter two, which together describe, seem paradigms of Educational Technology in intraeducacional dimension, with greater chances of positive results today.

We begin with the Evaluative Studies. In this group are those works that consider the possibilities of each individual technology and environment, which is analyzed and evaluated from an educational perspective. We can think of hundreds of research that has been done from this point of view during the twentieth century, from the first experiments which were carried out, for example, with the phonograph or the cinema in the early decades of the last century to very prevalent today on the personal computer, Internet, video games or virtual reality. All of them try to prove-or reject-the intrinsic value of each of these technologies for instructional media or work or in general education. Today one can say that these studies are only valid for the specific contexts in which they performed, monitored and carefully measuring all variables. Out of those specific frames, and of course the presence of other variables, the results can not be transferred, can not be generalized.

Today evaluative studies are rare, however long research models that were enjoyed great international prestige, which sought to establish ratings for the quality of education technology or medium. But at present it is impossible to speak of quality in relation to its most instructive-as ability may relate to their ability to transmit information, communication perspective, training characteristics, etc.

Another important constituent assembly Comparative Studies, analyzes between various media applied to any learning object. They have also been very common. The aim was to establish what technology or medium is more suitable for a particular area or subject, or to achieve certain objectives. Typically, the comparison between a novel and other traditional means, or within educational processes in which in one case a particular medium is used and in another it is dispensed, using a methodology based on exhibition techniques is performed. However, since the first experiences, and especially refer to those made by Freeman (1924) in the early twentieth century, it was shown that the most important thing is not the medium itself, but its use can be made of it, the student characteristics and related variables. Because basically the dialogue between social psychologists and educational media researchers, established a great interest in developing research on social media in the instructive power of them (Rogers, 1986) was compared, and if this is correspond to the presence and influence that each has on society.

Practically since the twenties, as we say, to the sixties of the twentieth century these studies were very common, but mostly we characterized by reliance on very fragile theoretical concepts, develop deficient experiments and not offer significant results (Hartley, 1974). These also led to contradictory or false results (Knowlton, 1964) which were collected in educational contexts producing harmful effects certainly are offered. Salomon (1981) found that the comparative failures of social media in education research were because they relied on false assumptions.

Thus, the initial question in comparative research on technology and media in educational settings, and many specialists still do, that is, which medium is the most effective for learning, no sense (Clark and Salomon, 1977) because it always depends on the context. There will be much more appropriate means to achieve certain objectives than others, no doubt, but will depend on where the learning of the whole teaching-learning process develops.

Much more interesting are the Intra-Medium (IM) and Aptitude Treatment Interaction (ATI), global model studies that we believe outweigh the flaws, but also have them, are of minor importance and are adapted to the features present ICT-of evaluation research and comparative research.

Back in the early sixties of the twentieth century was undoubtedly the failure of comparative studies, which led many researchers to seek new lines of research that will result from a more objective scientific perspective. Thus tried to describe the characteristics of each medium, i.e., to describe precisely its characteristics to determine their real possibilities for communication and interaction with students and with all variables in the teaching processes. Thus arose, as already noted, new models of educational technology research: the IM and ATI studies.

In the case of the former, the Intra-Medium studies, following the proposals offered by Salomon (1981) and Clark (1985), which research trying to measure all the variables to the medium itself is designed (presenting it as a constant), so it was possible to establish and compare different methods in which the same technology or means were used. From the affirmation of Salomon (1981) that the effectiveness of each medium depends on the nature of the instruction, the question was not now determine what is the most effective means, as in comparative studies, but what is the method most effective using that medium.

The various experiments conducted for ICT that have been made so far based on the methodology proposed by this model (although they have not been too many in number itself must emphasize quality) have been positive, and in some cases entirely satisfactory-like Lehrer and Randle (1987), for example, for computer-and are helping the development of teaching for the integration of new technologies as teaching resources strategies.

It implies, therefore, a model that we certainly interesting and can be very useful for the study of new technological realities, such as the phenomenon of social networking, Web 2.0 as a whole, the implementation of the company by m- learning in education, MOOC courses, etc..

The same could be said of the investigations Aptitude Treatment Interaction (ATI), which emerged in the early seventies of the last century as a new line of research which is mainly engaged in the constructivist paradigm. Hitherto, most studies were based on behavioral elements, but following the progress of psycho featured this time, it begins to be worked with cognitive theories that define learning as a process in which the learner makes a new integration knowledge of prior knowledge, and the means of communication (as currently ICTs) are presented suitable: they are capable of producing external stimuli that develop cognitive processes that can be led back to enhance learning.

In this situation into account different factors, all considered extremely important: student characteristics, abilities, prior knowledge, motivation, teaching methods, etc. (Clark and Surgrue, 1988 and 1990), highlighting the possibilities and types of interaction offered by various means to enhance learning. In this connection begin to conduct research that rely on the methodology, which then began to be called ATI, for the use of these teaching resources and was first defined by Cronbach and Snow (1977), although it was already outlined above by Parkhurst (1975). Recent examples of research using this model, and only as a sample button, we can mention those of Nouri and Shahid (2005), focusing on the use of PowerPoint in educational contexts, and Kieft, Rijlaarsdam and Van den Bergh (2008), investigating the importance of writing in learning.

The results that we provide in these studies is that there is no established or improved technologies or media, a priori, with greater learning opportunities within educational contexts, sophisticated and ideal for training-even if they are created or was designed for that purpose-will depend on many factors and variables.

The key therefore is to establish with absolute precision the object of our research, as this is key to the methodological development to follow. The whole methodology is always subject to what we want to investigate to what we investigate and how, indeed, should be investigated to obtain as much information as possible about it. It is essential, no doubt, in the research process in TEI.

4.5 Supraeducacional Educational Technology (TES)

4.5.1 General Features

We consider Supraeducacional Educational Technology (TES) is the main educational challenge existing today. Addressing the complex problems we face to only from an intraeducacional perspective is insufficient. The importance today of new technologies is such that it is necessary to train students in these elements as much impact and relevance in our society, if we really want to prepare citizens formed the current life. Education for ICT is necessary, not only has the presence of the same schooled as teaching assistants or productivity tools. It must be studied, analyzed and reviewed by the learner, because today would be part of a comprehensive education adapted to the needs of a world marked decidedly by computer and telemetric technologies, processes, information and communication. Not the user of a product in any way, as it is sometimes considered, is above all a human being who is part of a distinctly technological society. Integrating ICT becomes an essential educational challenge, in close connection with the other (political, economic, cultural, etc.) social systems. It's what we call a supraeducational dimension.

Educational research required for this is much more complicated and, if not properly focused, indefinitely. How to face the study of this complex object? The answer we offer is in relation to what we have presented above. It is appropriate to rely on assessments for long in this sense, and in the world of techno-media convergence (Gómez Galán 2003, 2007 and 2011), in which ICTs are now in essence means of communication or part of a single digital medium, which really demand perfectly valid and appropriate methodological frameworks for existing research models. And we mean, in particular, developed in media education.

When we talk about media education referring to teaching technologies and media to enable students to achieve the ability to analyze, use, and even issue in different ways messages produced by them. Itself mean a need to educate society to the media today. Apparently everyone knows something about the media. However, the knowledge of the consumer population and not as user (Hart, 1991). Really very few people are able to carry out a rigorous and critical analysis of media content. They know almost all basic communication characteristics of these media, codes, messages, and methods of influence. This leads, for example, for the consumption of its products is in most cases completely irrational. The population, in a comprehensive manner, is more demanding for any other product that produced by the media. And the main reason, among others, is that the company has not received adequate training in them, not as consumers, let alone as users. And in this context the role of the teacher as counselor is essential. Students should be able to take effect, critically, analyzing the information coming through ICT, powerful media whose influence is undeniable. As stated Jenkins, Clinton, Purushotma, Robison and Weigel (2006) is necessary for today's students acquire new skills to the new media culture that surrounds our society.

Of course we are talking about a challenge of education in relation to current social needs. All educational process that seeks the formation of active and critical citizens should influence today teach the correct use of ICT. In addition, critical media staffing means freedom to each individual, which is an essential function in any democracy. As a whole, critical literacy of the population can create a global awareness of action that makes people submit their own subjective judgments thus avoiding that are easily manipulated by powerful media groups (Masterman, 1990, 1991, 1995 and 2001, Giroux, 1992, Criticos, 2000, Buckingham, 2003 and 2005). Globally is what would create a functioning democracy. Some authors (Bereiter and Scardamalia, 1989) believe that precisely the formation of critical citizens is (should be) the primary function of all current education system (which must be structured to it), which, we add, could not be possible today without the challenge faced. Educational ICT research should especially affect it. These elements of our society, as we defend, are much more than mere tools that can be used as teaching resources. They constitute one of the bases of the XXI century.

4.5.2 The Media Education as Principal Paradigm TES

Research from this perspective, and because of its lack of tradition, has not been adequately covered by scientific and educational technologists. We could even take us back to the classical models of media education when were located in the field of analog still limited mass communication, but, however, no studies that we can offer an evolutionary and generally satisfactory vision. There are some attempts of interest, like Alvarado, Gutch and Wollen (1987), but that does not appear fully contextualized from a social economic perspective, policy, and it should definitely be present. We need a proper integrated methodological model for research in media education, and consequently ICT, flexible and satisfying, but also must take into account that possibly not meet before and identifiable linear process allowing better focus as accurate. Most likely that parallel models are necessary, suitable for molecular approaches. We have done so in previous research (Gómez Galán 1999, 2003 and 2011; Gómez Galán and Mateos, 2002a, 2002b and 2004).

Due to that, consider the state defend the issue in a comprehensive way, not essentially pedagogical, as were multiple perspectives from other fields (sociological, psychological, political science, semiotics, journalism, etc.) That have made inroads into the educational implications of the media and new technologies, and ultimately had greater influence and reception by specialists in the science of education. But the pedagogical dimension must always be present, otherwise the research processes incur dangerous reductionism by its consequences. For example, the paternalistic sense of teaching means that developed from the forties responded more to social and educational policies that claim, and had a major influence on subsequent studies. In the early eighties focused research studies preferably in nature artistic aesthetic that aimed to develop strategies for the student to acquire the ability to set quality criteria. It would be only from that decade when a new concept in the critical analysis from a representational and political perspective (media products as representations for certain purposes) replaces aesthetics (Masterman, 1998) emerges, and begins to be regarded seriously from the contributions of Hall (1977), the power of the media to disseminate ideologies. Research in media education begins to spread from the Anglo-Saxon countries, where he was born conceptually and methodologically, to the rest of Western educational systems, though a long way from the first.

At present, and in a particular way, we advocate the use of research models in the research process for ICT in educational settings, as well as teaching resources as an object of study (overcoming the division in both dimensions), because in definitive new technologies are now at the service of powerful mass media; are naturally means. And we've made both from a theoretical perspective and developing a model of field research, which in practice would summarize everything we've discussed in this chapter (Gómez Galán 1999 and 2003; Gómez Galán and Mateos, 2002a and 2002b).

Naturally methodological frameworks can be consistent with the objectives being pursued, and educational researcher may rely on those it deems most appropriate to your needs. Although as we have seen we propose a global model would also be possible to use molecular models from this perspective. For example, the educational and curricular approach in the sixties and seventies led to representational models advocating basically as we are exposing, for the need to study these important media for their ability to be creators and mediators of social knowledge, such that future citizens of any democratic society need to know how represent reality, how they do (techniques used) and what is the ideology that lies behind these representations, i.e., why not create them and extend. The best examples of this line of research we find, among others, in the models presented by specialists as Alvarado and Ferguson (1983) and Masterman (1990, 1991, 1995 and 1998). UNESCO itself (1984) had already been aware of the importance of studies in this regard.

The Media Education (and ICT, we consider unified for all the reasons we have presented), from this new research paradigm, acquires decisive importance in the formation of future citizens (in no way is a training supplement and / or cultural). This new lens is so significant that could change, including the proper consideration of the nature of the educational curriculum. And the challenge today is undeniable, increasingly emphasized by the great development experienced ICT in recent years, and the convergence to the digital world has led to communication processes. Together with a second youth of traditional media (also enhanced by its development from new technologies), we are now witnessing the explosion in the use of Internet and Web 2.0, inclusive systems of multimedia technologies, extraordinary power development and possibilities of informatics and telematics systems, mass storage and cloud (cloud), video games, virtual reality, mobile unlimited possibilities, tablets, etc.., are examples of technologies that are beginning to take center stage in this and dominate the morning, and allow the media to provide information quality (and quantity) never ever, with all that this implies in terms of their influence on society and, of course, education. The way to respond to all this will be only one: the curricular integration of ICT in a serious, precise and rigorous way.

PART II

Models and Experiences in Higher Educational Research

5

Measurement of Satisfaction with the Online and Classroom Teaching Methodologies of Graduate Students: A Descriptive Research

Mylord Reyes Tosta

Scientific Research Services, LLC, MI, United States

5.1 Introduction

This chapter was written with the purpose to inform the entire school community and other interested individuals of the findings from a study comparing online courses with classroom courses. These results on satisfaction and academic achievement may be the starting point for making decisions aimed at improving academic structures and processes of graduate schools. The growth of higher education through online courses in Puerto Rico is remarkable (Núñez, 2009). Therefore, it was necessary to know its status as there are many technological devices involved in the processes and they change very often. This is a recent perspective of the teaching-learning process that has important implications for attitudes and academic performance of students by being a method that uses new technologies and breaks with the uses, paradigms, and ideologies that precede it.

Taking this into account, this study scientifically examined what was the degree of satisfaction and academic achievement of students who participated in online courses while they are still immersed in traditional education. In addition, an instrument was developed and validated to collect the information according to the nature of education in Puerto Rico. The validation process provided valuable instrument to the study. Different theories were strongly influenced by research in the field of business (Pichardo, García, De la Fuente & Justicia, 2007), and this fact should provoke greater interest in knowing the satisfaction university's students as a variable to consider in improving the quality of the services of higher education institutions.

5.2 Objectives

The following were the objectives of the study:

1. To design and validate a data collection instrument to determine student's satisfaction with the online and classroom courses.
2. To measure student's satisfaction with the online and classroom courses.
3. To identify the difference in student's satisfaction between the online courses and classroom courses.
4. To identify the difference in student's academic achievement between the online courses and classroom courses.
5. To describe the gender differences found in the academic achievement of students within each of the two types of teaching methodologies.
6. To describe the gender differences found in student's satisfaction within each of the two types of teaching methodologies.

5.3 Design

This study used an exploratory design (Figure 5.1). According to Waters (2007), exploratory research provides an understanding of a phenomenon or situation. It is a type of research conducted when a problem has not been clearly defined and it helps to determine the best research design, method of data collection, and selection of subjects. It allows definitive conclusions only with extreme caution. DJS Research Ltd. (2010) points out that, unlike descriptive research, the exploratory research is used primarily to obtain a deeper understanding of something. Its design is more flexible and dynamic than descriptive research. In general, it research trends, identifies potential relationships between variables and set the tone for subsequent research more rigorous.

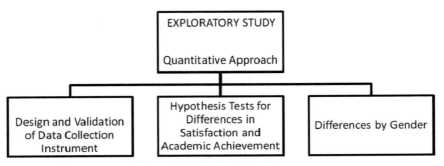

Figure 5.1 Study design diagram.

5.4 Sample

This study was performed in 2011 and a convenience sample was used to identify the sections for the two teaching methodologies (online courses and classroom courses). The sample for the two teaching methodologies consisted of 74 graduate students at the master level who enrolled in Human Resources Management (HURM 710) and Organizational Behavior (MANA 501) courses (Table 5.1).

Table 5.1 Distribution of study sample by course and gender

| | Courses | | | | | | | | | |
| | HURM 710 | | MANA 501 | | HURM 710 | | MANA 501 | | | |
Gender	Online	%	Online	%	Classroom	%	Classroom	%	Totals	%
Male	7	41%	3	21%	6	30%	8	35%	24	32%
Female	10	59%	11	79%	14	70%	15	65%	50	68%
Total	17	100%	14	100%	20	100%	23	100%	74	100%

5.5 Data Collection Instrument

The data collection instrument designed for the survey was a questionnaire (Table 5.2) and it consisted of 48 items using a 5-point Likert scale, where 5 = Very Satisfied (VS), 4 = Satisfied (S), 3 = Somewhat Satisfied (SS), 2 = Dissatisfied (D), 1 = Very Dissatisfied (VD) and Not Applicable (N/A). A panel of experts was used for the validation of this instrument and a Cronbach's alpha coefficient was calculated after its administration to measure its reliability (quantitative indicator of content validity). All students' academic achievement measurement was completed based on the final grades obtained by the students in each course. The final grades were offered by the students in the survey. The survey was anonymous and personal and/or sensitive data was not requested.

SATISFACTION SURVEY WITHT EACHING METHODOLOGY
Part I: General Information

Instructions: Please enter the date you completed this survey and mark in the blank if you are enrolled in an online course or classroom course and your gender (male or female). Also, write the name of the course and check the grade obtained.

Date: _____ Online course: _____ Classroom course: _____
Gender: _____ Male _____ Female
Course name or code: _____
Course grade: A ___, B ___, C___, D___, F___

Part II: Measurement of Satisfaction with Teaching Methodology

Instructions: Below is a list of independent variables that measure your satisfaction with the teaching methodology. Please read carefully each one and mark (×, ✓) the option that best describes your opinion using the following scale:
Scale: 5 = Very satisfied (VS), 4 = Satisfied (S), 3 = Somewhat Satisfied (SS), 2 = Dissatisfied (D), 1 = Very Dissatisfied (VD) and Not Applicable (N/A).

Table 5.2 Satisfaction survey with teaching methodology

Variables to Measure Satisfaction	5(VS)	4(S)	3(SS)	2(D)	1(VD)	N/A
1. Student and instructor interaction						
2. Faculty feedback to student progress						
3. Time to get the course						
4. Homework delivery time						
5. Course period						
6. Professor response to questions from the students						
7. Student-to-student collaborations						
8. Communication between teacher and student						
9. Teacher assistance to solve student problems						
10. Group work results						
11. Student's learning styles considered when designing course strategies						
12. Techniques used by the teacher to teach the course						
13. Integration of theory with practice						
14. Activities used to support the practice						
15. Activities used in the course according to the objectives						
16. Teacher content mastery						
17. Updating the content of the course by the teacher						
18. Course management for self-motivation of the students						
19. Availability of the course syllabus provided by Professor						
20. Teacher's support on time outside the classroom						

(Continued)

Table 5.2 Continued

Variables to Measure Satisfaction	5(VS)	4(S)	3(SS)	2(D)	1(VD)	N/A
21. Technology according to the course objectives						
22. Initial instructions on the use of technology						
23. Information technology resources available						
24. Teacher technology mastery						
25. Technology resources updating						
26. Registration services						
27. Payments services						
28. Financial aid services						
29. Programmatic services						
30. Updated library services facilities and physical information resources						
31. Updated library services related to virtual information resources						
32. Library services schedule						
33. Library environment						
34. Library services' customer support quality						
35. The course curriculum provides achievable objectives						
36. Objectives of the course curriculum are aligned to its content						
37. Fulfillment of course objectives						
38. Upgrading of course curriculum						
39. Effectiveness of student's formative evaluation						
40. Effectiveness student's summative evaluation						
41. Techniques used in the evaluation according to course content						
42. Equivalence in the qualitative and quantitative student's evaluation						
43. Student's assessment feedback						
44. Physical facilities accessibility						
45. Facilities environment						
46. Facilities food availability						
47. Facilities health services						
48. Facilities maintenance						

5.6 Data Analysis

Descriptive and inferential statistics were used for data analysis. The statistics used to answer all research questions are presented below:

1. Frequency distribution table for both satisfaction survey and the students' grades for each of the courses.

2. Table of descriptive statistics for the satisfaction survey and the grades of the students (mean and standard deviation). These were used to enable the reader to visualize and better understand the results showed by the inferential statistics.
3. Cronbach's alpha coefficient to measure the reliability of the instrument.
4. Nonparametric statistics (Mann Whitney U test) for independent samples were used at a 0.05 significance level. They were used to establish whether there was a statistically significant difference between the two methodologies related to satisfaction (Jandaghi & Matin, 2009).
5. A hypothesis test for two independent means was performed, using the Mann Whitney U Test, to establish whether there was a statistically significant difference between students' grades for both types of teaching methodologies.
6. A hypothesis test for two independent means was performed, using the Mann Whitney U Test, to determine whether there was statistically significant gender difference in students' grades by gender within each of the two types of teaching methodologies.
7. A hypothesis test for two independent means was performed, using the Mann Whitney U test, to establish whether there was a statistically significant difference in student satisfaction by gender within each of the two types of teaching methodologies.

The Kolmogorov-Smirnov/Shapiro-Wilk test was used for testing normality of the data and, to establish the differences by group, the nonparametric tests Mann-Whitney U and Kruskal Wallis were used because not all data fulfilled the assumption of normality, except one of the variables of satisfaction. In this last case, the t test was used. An alpha of .05 (Type I error) was used in all hypotheses tests of this study. The Statistical Package for Social Science (SPSS) was used for statistical calculations.

5.7 Answers to Research Questions

Based on the results obtained from the data analysis of this study, the answers to the five research questions are presented below.

1. What are the psychometric properties of the data collection instrument? The psychometric properties of the data collection instrument were established by the reliability given by Cronbach's alpha coefficient, which measured the internal consistency of the scale. That is, the linear correlation between the item and total score was established. The Cronbach's

alpha coefficient was .984 for both teaching methodologies according to participants' responses. The Cronbach's alpha coefficient contributed to the high degree of reliability to the data collection instrument. The results compare favorably with the statement of Kim & Feldt (2010), when they indicated that a coefficient higher than .70 provides an acceptable reliability to the instrument.

2. Is there a statistically significant difference in student's satisfaction between the methodologies of online courses and classroom courses?
   The hypothesis test of Mann Whitney U showed a result of $p = .295$ ($p > .05$). Therefore, the null hypothesis was retained and the alternate hypothesis was rejected. There was not a statistically significant difference in student's satisfaction between the methodologies of online courses and classroom courses. The nonparametric test of Mann Whitney U was used because the data did not show normality (Table 5.3).

Table 5.3   Student's satisfaction normality test by methodology

Methodology	Normality Tests of Kolmogorov-Smirnov/Shapiro-Wilk
Online Courses	.012
Classroom Courses	.003

3. Is there a statistically significant difference in academic achievement of the students between the modalities of online courses and classroom courses?
   The hypothesis test of Mann Whitney U showed a result of $p = .774$ ($p > .05$). Therefore, the null hypothesis was retained and the alternate hypothesis was rejected. There was not a statistically significant difference in the academic achievement of the students between the methodologies of online courses and classroom courses (Table 5.4).

Table 5.4   Average, standard deviation, and hypothesis test of academic achievement by methodology

Online Courses			Classroom Courses			Hypothesis Test
n	*M*	*SD*	*n*	*M*	*SD*	*Significance*
31	3.74	0.44	43	3.67	0.47	.774

4. Is there a statistically significant difference in academic achievement by gender within each of the two teaching methodologies?
   The hypothesis test of Mann Whitney U showed a result of $p = .533$ ($p > .05$) in the comparison of the academic achievement by gender in the

online methodology. Therefore, the null hypothesis was retained and the alternate hypothesis was rejected. There was not statistically significant difference in academic achievement of the students by gender for the methodology of online courses. Moreover, the hypothesis test of Mann Whitney U showed a result of $p = .699$ ($p > .05$) in the comparison of the academic achievement by gender in the classroom methodology. Therefore, the null hypothesis was retained and the alternate hypothesis was rejected. There was not statistically significant difference in academic achievement of the students by gender in the methodology of classroom courses (Table 5.5).

Table 5.5 Average, standard deviation, and hypothesis test of academic achievement by methodology and by gender

Methodology	Male			Female			Hypothesis Test
	n	M	DS	N	M	DS	*Significance*
Online Courses	10	3.80	0.42	21	3.71	0.46	.533
Classroom Courses	14	3.64	0.50	29	3.69	9.47	.699

5. Is there a statistically significant difference in student's satisfaction by gender within each of the two teaching methodologies?

   The hypothesis test of Mann Whitney U showed a result of $p = .118$ ($p > .05$) in the comparison of satisfaction by gender for the online courses. Therefore, the null hypothesis was retained and the alternate hypothesis was rejected. There was not a statistically significant difference in student's satisfaction by gender in the methodology of online courses. Moreover, the hypothesis test of Mann Whitney U showed a result of $p = .136$ ($p > .05$) in the comparison of student's satisfaction by gender in the methodology of classroom courses. Therefore, the null hypothesis was retained and the alternate hypothesis was rejected. There was not a statistically significant difference in the student's satisfaction by gender in the methodology of classroom courses (Table 5.6).

Table 5.6 Results of hypothesis tests of total satisfaction by gender of online and classroom courses

	Hypothesis Test
	Significance
Satisfaction by Gender for Online Courses	.118
Satisfaction by Gender for Classroom Courses	.136

5.8 Discussion and Educational Implications

The results of this study coincided with the study of Alonso et al. (2009), whose purpose was to determine if the results of distance learning using a particular teaching methodology are comparable with the traditional teaching methodology (face to face). The results indicated that the degrees and levels of satisfaction were similar for the students studying through distance and traditional learning. Moreover, in the study of Markauskaite (2006), who sought gender differences in three satisfaction variables in graduate students candidates to be teachers, no significant differences were found between female and male on experience and literacy related to information and communications technology (ICT). This gender comparison of Markauskaite coincided with the present study, but did not match the other variables where significant differences were found. However, in the study of Dolan (2008), no statistically significant differences were found in any of the 14 variables of satisfaction with the course in the comparison between the online methodology and the classroom methodology. Therefore, the results of Dolan's study concur with the results of this study.

These results are in accord with those obtained in a study by Liedholm and Brown (2001), where no statistically significant difference was found between men and women from a group of college students taking a graduate level course online. Moreover, in the classroom course, the findings were not consistent because the study of Liedholm and Brown did find a significant difference in men with a higher performance. They stated that this finding was consistent with previous research showing that many women do better in online courses than in classroom courses. The same trend was found in the results of this study.

Also, the results of this study agree with those of Stewart, Choi & Mallery (2010) who found similar results in the two teaching methodologies of their study, whose purpose was to investigate whether the academic achievement of students, measured by the grades, differ between the online course and the traditional course. The results of the study of Choi & Johnson (2005) were similar and did not find a statistically significant difference between the two methodologies, where the mean for the online course was 20.06 and the mean for the traditional course was 19.88. The study by Ellis (2005) also found results suggesting that students in the traditional classroom and virtual classroom executed at a similar level.

Very similar results to this study were those of Gratton & Stanley (2009), who investigated the gender difference between the two teaching methodologies. The results did not show statistically significant differences in both methodologies of teaching and in the gender comparison. However, the raw data reflected a higher average in online learning and for males.

A similar scenario was the study of Barrett & Lally (1999), which explored gender differences in students in online courses at the graduate level in the use of communication through computers in a specific learning context. As for the grades, the performance of females and males was similar, but their social and interactive behavior differed significantly. While in the study by Price (2006), in terms of academic achievement in online courses, there were statistically significant differences between females and males, higher for females. As for the academic achievement of traditional courses, there was not a statistically significant difference between females and males.

But beyond this discussion about the results of comparable studies, this study on the comparison of teaching by two courses delivered through online and traditional methodologies to graduate students, using the results on levels of student's satisfaction and grades from both groups in the same course, produced straighten data on the attitude and cognitive aspects. A very important fact about this study is its contribution to strengthen the procedure used toward the search for more comprehensive results, because an instrument to measure student satisfaction was created and validated.

This instrument consisted of a questionnaire using a Likert scale to measure 48 independent variables (causes), which affect satisfaction (the dependent variable or effect). The instrument had a holistic design that included both the online and classroom methodologies. It was validated and its reliability was established, therefore, it can be used by other researchers, teachers, and the educational community.

The results obtained in this study showed that there were not statistically significant differences by type of teaching and gender for the satisfaction and academic achievement variables. This new scenario now known, offers more information to make better decisions to maximize the quality of educational processes. Another contribution evidenced by the results of this study was the possibility to answer any questions that many students and education professionals have about the effectiveness of the information and communications technology. Also, to have a clearer vision of the implications and the cost of technology on the productivity of higher education, in terms of satisfaction and academic achievement, is a benefit because it allows improving the strategies used in the teaching learning processes. Additionally, it allows creating expectations about the direction that technological advances in the field of education could have. Therefore, more research is needed to know if some of the paradigms that have survived in traditional education are ready for change.

This paradigmatic, structural, procedural, and philosophical shift could be mainly due to rapid advances in information and communications technology. These technological advances have improved the quality of processes and outcomes, have saved time and effort because the devices are more efficient and have removed physical spaces because there is no distance, in terms of communication. Thus, with the acceptance of these benefits, there is less resistance to change every day.

5.9 Conclusions

The results of this study suggested that a well-designed online course can be as efficient and effective as a well-designed course in the traditional methodology. Also, it was inferred that the fact that the online methodology using the Blackboard platform was more recent and unknown to many, does not mean students cannot achieve high levels of academic achievement and satisfaction.

Moreover, a valuable contribution of this study was the design of a data collection instrument that was validated by a panel of experts and its reliability was established through the Cronbach's alpha coefficient. A correlation coefficient of .984 contributed to the high degree of reliability of the data collection instrument. The data collection instrument consisted of a survey using a Likert scale to measure 48 independent variables (causes), which affect satisfaction, the dependent variable (effect). The data collection instrument considered the most important components of an institution of higher education and is of great relevance for students. It was evaluated by participants to determine their satisfaction, aspects directly related to teacher effectiveness, student performance, technology, administrative and program services, physical and virtual library services, course curriculum, and physical facilities. In addition, the instrument provided the student's performance on the course and it was applied to both, students of online courses and classroom courses.

This study did not find statistically significant differences in the satisfaction and academic achievement of the participants according to the responses obtained through the survey. In the hypothesis tests performed between the online and classroom methodologies, there was not statistically significant difference in satisfaction and academic achievement. The same results were obtained in the hypothesis tests performed by gender, comparisons within online courses and within classroom courses. That is, it did not find statistically significant differences in overall satisfaction and academic achievement by gender in any of the two types of teaching methodologies.

5.10 Recommendations

Based on the findings of this study on measuring satisfaction and academic achievement of students and its comparison between teaching methodologies, including gender, recommendations were made for graduate schools with online and classroom teaching methods. For managers and administrators, it was recommended to include more research courses in academic programs, expanding the fields of study that include online courses and finding strategies to motivate students to learn this methodology. Also, encourage faculty to engage in research proposals.

For the faculty, it was recommended to encourage students to collaborate on research, highlighting the importance for themselves as students and for educational community. Also, for the research and evaluation courses, to present as a topic of discussion, the data collection instrument produced in this study to analyze it in all its parts. In addition, to post in the Internet the results of past and future research completed in the courses.

For other researchers, it is recommended to conduct studies on academic achievement and satisfaction in both teaching methodologies at the undergraduate level, including other public or private universities, and a qualitative approach regarding its advantages and disadvantages. Also, it is recommended to investigate the factors influencing student's academic achievement and satisfaction for online and classroom methodologies. In addition, based on the results of this study, it is recommended to conduct studies using a quantitative design with treatment in experimental group and with a control group.

It is recommended that educational researchers use the reliable and validated data collection instrument produced in this study. Given the features, design and scope of this instrument, researchers can use it for public and private universities, for undergraduate and graduate levels, and for any course and academic area. It is easily adaptable to suit the needs of each study and applies to both types of teaching. In general, the components of this instrument cover the most important of all academic structure, known until now, allowing the researcher to obtain necessary and sufficient data to draw conclusions.

Given the importance of data collection instruments for achieving the goals of research, it is encouraged to use other methods to measure the reliability of the instruments, and make comparisons between them. Moreover, it is recommended to Puerto Rican researchers, to publish the results of their studies. In the literature review conducted for this study was not possible to find data related to the research problem within the educational environment of Puerto Rico.

6

Measurement Adult College Student Experience: A Phenomenological Analysis

Zobeida González-Raimundí

Scientific Research Services, LLC, MI, United States

6.1 Introduction

Many social and demographic changes had occurred in Puerto Rico since the early 80's. Due to those changes there is an increasing demand for mature and senior adults with a college degree. In response to that demand, many universities developed academic programs focused on the adult student. But those programs did not take in consideration if they appealed or not to the population of adult student. So, it soon was evident that the new curricula and complimentary services offered were not in accord with the reality of the students they were intended for. That discordance became the problem presented in this dissertation.

One objective of the study was to bring recommendations based on their outcomes that can help to improve and transform nontraditional education programs in Puerto Rico. This research, through the phenomenological prism, allows us to understand the experience of students, 50 or over, who enter a university to pursue a bachelor's degree, through nontraditional academic program intended for adults.

Education is a strongly tied to the psychological, social and moral development of all human beings. This process, continuous and permanent, is not exclusive of determined stage of growth or development. It attains to all individuals, independently of the age group; one of these groups is that of mature and senior adults. In accord with this postulate, an increase in the number of mature and senior adults entering or reentering universities is observed in Puerto Rico and in other countries. This phenomenom had the caught interest in all societies sectors: government, organizations and individuals, throughout the world (Longworth, 2003).

Entering this phenomenom and going deep into aspects such as feelings, opinions, perceptions and ideas that propitiate, enrich the educational context in which it happens. It also enhances its perspectives and strengtheners the teaching-learning processes of the group mentioned. It is considered that discovering the essence of this phenomenom and acting in accord of its demands, contributes to enrich the work of the people immersed in those processes.

6.2 Methods

The study took place during the first semester of 2012–2013 academic year. The qualitative methodology was employed, in the phenomenological interpretative (hermeneutical) mode, based on the inductive research model and framed in the constructive paradigm. Three universities were selected from a list of higher education institutions in Puerto Rico, provided by the Carnegie Institute; these universities had academic programs focused on the adult student. The sample for this study: nine students (informants)-three from each university-, 50 or over, was selected by invitation, through a memorandum posted in the bulletin boards of the campuses.

All the informants were required to complete a brief questionnaire about social, academic and demographical data, in addition to answering 13 guide questions in individualized interviews, for approximately 90 minutes. At the end of each interview, the researcher registered the observations about the informants and her/his environments during the process. The processing of the obtained data was made in accordance with the Model suggested by Wolcott (1994): description, analysis and interpretation.

6.3 Outcomes

The first stage of data processing allowed the description of the outcomes, some of whom are exposed as follows. One of those was that all of the informants expressed having felt fear, doubt and/or anxiety prior to deciding if entering or going back to university. They were worried about the reactions of the people knowing that they will be college students. They also worried thinking that they could have lost their capacity to learn. At the moment of the interviews, they changed these feelings for the assurance that in college, all students are considered equals in terms of the objectives they persuade, no matter age, as long as they are confident in their capacity and the wish to complete their degree.

The study also showed that the informants perceived the academic goal as the peak of their self-actualization (Maslow, 1991): to complete the degree that cannot be completed during many years, mainly because of the lack of economic resources. They felt authorized even though had not finished their degrees. The informants considered that it was fundamental to give the best of theirselves in the university experience. That way, they expressed with emphasis how significant was for them to maintain an excellent academic average. The answers also exposed that the time assigned for each course was too short to cover the topics and to receive individualized attention and that professors did not have offices or office hours to meet with adult students. The informants considered that these factors negatively affect their academic performance.

The findings merit to be analyzed based on the review of literature about adult education and the professional experience of the researcher. As to the feelings: disorientation and/or anxiety, Knowles (Knowles, Holton, Swanson, 2001) establishes that those feelings are rooted in the rejection to which these students are exposed; that rejection coming from their younger classmates. Knowles also points out that as a consequence of the adult experience, in front of the limited experience of the younger students, they think that the young ones only want to attract the attention of her/his companions. The informants considered that being university students could caused them: discomfort with relatives and friends, feeling of rejection by the younger companions, fear of not being understood and thought that they could lack the capacity to compete with the young students. In terms of the goal of self-actualization, Maslow (1991, p. 32) establish the importance of it: "That necessity can be called self-actualization . . . the wish to be as much as one can be, in accord to her/his idiosyncrasy; to attain to be what everyone is capable to be".

As to the answers in which some informants expressed the factors that affect their academic performance, they confirm the research problem. The researcher understand that those factors should be attended with urgency, so that adult programs accomplished their mission of providing that population with the learning experiences that will capacitate them as professionals as well as human beings. It is precise to consider that adult programs are structured in fast modes and owning that the hours assigned for each course, professors do not have enough time to give students individualized help.

In this analysis there should be considered a period to bring students the individualized attention required by them, particularly if professors do not have offices hours to give that help. It is also considered that during the adjustment process to university life is very important to bring the adult student the necessary time to adopt to her/his new life style. In addition, it is recommended the establishing of a link office in the campus, which service hours are in accord to the needs of adult students, so they can communicate with professors, as well as having friendly meetings with their peers.

6.4 Conclusions

The informants directed their answers, essentially, to one of three aspects in terms of their university experience: the attaining of their academic goal of completing the bachelor degree, as the peak of their self-actualization, the development of critical thought and their optimal development as a human being.

Through the exposition of the answers to the research questions, the research problem is exposed: Even though the university institutions in Puerto Rico establish academic nontraditional programs targeted for adults, these are developed without a profound exploration of aspects such as: feelings, life experiences, perceptions, believes, descriptions and opinions about what the experience of being university students means to them, reason for which there are curriculum modalities and complementary services for the adult student that do not respond to the reality that belongs to the experience. In this aspect, the study sustains that the findings of various investigations, done in Puerto Rico and abroad (Bowman & Burden, 2002; Fernández, 2008; Rivera, 2008).

The following was the central question of the research: how do you explain, from a phenomenological perspective, the experience of the university student 50 years of age or over, that are enrolled in a bachelor's degree in Puerto Rico, through the nontraditional programs targeted for adults? From the phenomenological perspective, this experience is considered as transcendental; to expose the informant to the exercise of giving ample responses and profound guide questions to the interview, it guided them to go into the reality of the phenomenon. In this sense and in a natural way, it was observed that there was an extraordinary change on the core of each adult, through the inductive reasoning that was employed when the answers were analyzed, and that they exhibited the maturity of the experiences of the university students had represents in all aspects of their lives.

The findings of the studio sustained all the premises of Knowles's Andragogic Model (Knowles et al., 2001), as for the hierarchy of the Necessities of the Human Being, postulated by Maslow (1991), of which the highest was auto-realization. The mature adults and older adults in Puerto Rico, who took the decision to enroll or return to the university with the goal of obtaining a bachelor's degree, as the most elevated aspiration, they demand and deserve to have opportunities and the conditions provided to help them reach it.

Focused on the results of this research, there are a variety of recommendations, some of which are mentioned as follows. One of them is to modify, in universities in Puerto Rico, the oral and written vocabulary used in the promotional signs and in the efforts to have people enroll in the university. Through these, conscious or unconsciously, there is special focus on youth and the presence, which is growing rapidly, of the adult as a university student is ignored. Through the special attention given to adults, the institutions would be able to integrate them to the rest of the alumni and expose the marked presence of the people of a mature age to the rest of the university's population. The researcher also recommends that the institutions of higher education in Puerto Rico consider among their strategic plans, the revision and the strengthening of the curriculum targeted to adults, in accordance to their demands and needs, in which the results of studies like this should be considered. She recommended that when the academic programs for adults refocus, they should consider as a cornerstone, the life experiences of the mature adult and the advanced adult. A practical way of taking advantage of this experience would be through a mentoring program, through which the young students can receive the supplies with the experience of the adults, related to the content of the courses and with others aspects of university life.

7

Equality Studies in Higher Education: A Model of Qualitative Research Method

Mariwilda Padilla Díaz[1] and José Gómez Galán[1,2]

[1]Metropolitan University, PR, United States
[2]Metropolitan University, PR, United States,
and University of Extremadura, Spain

7.1 Introduction

The topic of gender equity has been thoroughly discussed in the social sciences and in education (Martínez Ramos & Tamargo López, 2003; Salomone, 2007; Chisamya, DeJaeghere, Kendall & Khan, 2012; Sinnes, 2012; Weaver-Hightower & Skelton, 2013; Klein, Richardson, Grayson, Fox, Kramarae, Pollard & Dwyer, 2014). Historically, the attention dedicated to the study of the topic from a scientific and academic standpoint traces back to the 1960s, to the fight for the recognition of civil rights in the United States (Corey, 2013). In this sense, Brown (2006) maintained that the feminist movement organized during this decade was characterized by a widespread struggle between groups that demanded equal rights facing racial and gender discrimination, as well as peace in the context of condemnation of the Vietnam War. According to this author, the demands of these groups, representative of a wide-ranging social movement, gave way to the advocacy of a relation of equity between the genders. A change in social thinking was initiated, promoted by these feminist groups, articulated in two important directions: (1) the reformation of laws against discrimination against women and (2) ideologies directed towards the transformation of an androcentric worldview, that is to say, focused on inequality. The predominant discourse was directed essentially towards absolute equality between men and women. In this vision, the understanding of gender equity was focused almost exclusively in the discourse of equality (Salomone, 2007). The author maintained that the discourse of equality flowed during this historical moment between a model of similarities that demanded a justice focused on normative masculinity, and homogeneity and equality of rights.

Since then, the demand of justice centered on normative masculinity has subjected to much criticism, particularly from feminists with postmodernist worldviews. For instance, some argue that the discourse of equality should be set forth in the context of *differences*, since there is no such thing as a coherent and unified feminine subject (Weiler, 2001). The postmodernist approach, for example, argues that there are distinctive ways of being a woman, for which reason gender includes subjective constructions which themselves constitute the plurality of being a woman (Montrero, 2006). Although this article does not circumscribe itself exclusively to postmodernism, we will be able to see that the participants in the current investigation are centered on the recognition of differences.

The discourse of equality of rights has captured global attention, for which reason we will introduce some initiatives of great educational impact. In 1995 the fourth World Conference on Women took place in Beijing, China (Grunberg, 2004). This conference had a momentous reach in educational politics on gender and educational equality. They were aimed towards access by and non-discrimination against women in educational settings, increasing retention rates for girls and promoting discrimination-free educational environments. In 1998, the UNESCO celebrated the World Conference on Higher Education, in which strategies for the development of women in higher education were proposed. In 2000, the United Nations revised the development goals for the millennium, with particular attention to goal number two, focused on promoting gender equity and empowerment by women (Moletsane, 2005).

In April 2000, the World Education Forum concluded that gender discrimination continues in current society, and that hundreds of millions of people around the world are illiterate (Moletsane, 2005). This forum has established that education is fundamental for economic prosperity and social development, as well as to eradicate poverty at a global level. Additionally, the European Union named 2007 the European Year of Equal Opportunities for All: Towards a Fair Society (Jiménez Fernández, 2011). The European Center for Higher Education (CEPES) is an organism dedicated to supporting the movement of women in Central and Eastern Europe (Grunberg, 2004). According to the author, CEPES has achieved great advances, but is limited due to the incompatibility of educational models from one region to the next.

In the 90s in Latin America, the Economics Committee for Latin America and the Caribbean promoted an educational reform for the transformation of inequity in education (Fuentes, 2006). After two decades of implementation, it was found that the system led simultaneously to integration and to exclusion.

This data leads one to infer that in Latin America, gender equity has not yet constituted a priority in educational systems. Among other aspects, it was found that there is a great dismantling and lag in education facing the demands of the globalized world, and that sexist patterns are constantly reproduced in this region of the world.

In parallel, in a sociological study carried out in Spain, and in which we participated, it was found that kids and youths of this region still perpetuate traditional gender stereotypes, reproducing sexist patterns (González Pozuelo, Gómez Galán, Pérez Rubio, Blanco, Rumbao and Navareño, 2009). It was determined that both in the domestic setting and in the respective educational scenario, sexist ideologies prevail. As we can see, the implications on the topic of gender are varied and of great impact in our current society; hence the importance of the role of education for equality.

Velasco (2007) proposed that it is necessary to train teachers to consolidate a culture of equality in school, and to eliminate discriminatory approaches in curricula. It is also vitally important to promote the capacity for reflexive academic spaces centered on gender equity and on deploying programs directed towards orientation, gender studies and investigations. Jiménez Fernández (2011) proposed that education for equity is a complex process that requires strategies centered on non-discrimination and gender consciousness. To achieve this, suggests the author, a critical analysis is required on what is included and what is excluded in education, and on how topics dealing primarily with gender are approached. She added that it has been proven that environments free of sexist discrimination have positive effects on enrollment and on school retention rates.

7.2 Equity and Education: Areas of Study

But the social and educatives areas of influence of this issue are very broad (Inglehart & Norris, 2003; Wrigley, 2003; Unterhalter, 2005; Aikman & Unterhalter, 2005; Pateman & Grosz, 2013; Connell, 2014; Banks, 2015). We should start with the fact that inequality between men and women is not a problem of the past, in no way whatsoever. Although throughout history, and since we have sources of social knowledge, discrimination against women has been a constant in practically all ages and cultures, the problem is still current and completely prevalent. And it is not limited to the economic, work and professional areas: it is produced in all facets of the human being, and it generates some ignominious phenomena in our species, like gender violence, with all the impact it has and whose eradication is one of the most important educational and social challenges.

Although it is true that the main organizations and international documents indicate that the term "gender" implies both sexes, and it is thus described by the UNO (1986, 1992, 2008), Human Rights Watch (2001) and the Statute of the International Criminal Court, also called the Rome Statute, in its article 7.3 (ECPI, 1998), it is also true that women are the main victims in most acts produced in the context of gender violence. These represent a grave violation of universal rights protected by international conventions and organisms, and in all of its manifestations, they are considered illegal and criminal acts in the majority of national policies and legislations (UNFRA, 2012); unfortunately, in more than a few countries, this discrimination is directly defended by the state. There are so many atrocities in this field, systematically carried out in many countries, such as general psychological, physical and sexual aggression, rapes, ablations, feminine infanticide, human trafficking and prostitution, labor exploitation and many more, that one sometimes gets the impression that the problem in the Western, developed world is not so important. However, actions such as work discrimination, sexist acts or, fundamentally, domestic violence, are completely universal, and it is not possible to pinpoint the influence of one specific variable (economic situation, culture, religion, etcetera) among its causes.

Both in Latin America and in Europe, for example, domestic violence is a first-level problem. In the first case, according to the study of the Observatory for Gender Equality of Latin America and the Caribbean (CEPAL, 2013), carried out among Latin American countries, and in the second case, in a very recent study carried out in the set of European countries by the European Union Agency for Fundamental Rights (FRA, 2014), the percentage of psychological or physical violence reported by women represented scandalous percentages. Even in countries as developed as Sweden, Finland or Denmark, the percentage of women who claimed to have suffered some type of abuse rose as high as 28%, 30% and 32%, respectively. In the set of the European Union, around 12% of those interviewed indicated they had suffered some degree of sexual aggression or incidence at the hands of an adult before they were 15 years old, which amount to 21 million women in this region, which we should remember is among the most advanced in the world and with the most active policies for equality. Additionally, the results of this report reveal that 30% of the women that have been victims of sexual aggression at the hands of their current or past partners had also suffered sexual violence during their childhood, while 10% of women who had not been victims of sexual aggression in their current or past relationships indicated that they had experiences of sexual violence in their childhood; a clearly bleak picture.

Although comparison is difficult as different methodologies are employed, other studies carried out in other developed regions of the planet also show results in which domestic violence appears as a very important social problem. In the case of the United States, this can be seen in the study carried out by Black, Basile, Breiding, Smith, Walters, Merrick, Chen and Stevens (2011), in a sample inclusive of all 50 states of the country. Worth a mention is the study carried out a decade ago by the World Health Organization (2005). This is decisive in Third World countries, where policies that aim towards equality not only offer educational and economic benefits in the fight against poverty, but are also fundamental for health (Tyer-Violay and Cesario, 2010), even representing a marked reduction in birth mortality rates for women (Singh, Bloom, Haney, Olorunsaiye and Brodish, 2012).

The social dimensions of influence in the issue presented are multiple, severe and of worldwide reach. Discrimination in all of its manifestations has very negative effects on society and puts our evolution in serious question (Seguino, 2000; Grusky & Weisshaar, 2001; Branisa, Klasen & Ziegler, 2013; Hurst, 2015; Cavalcanti & Tavares, 2016). It has consequences so grave that it is vital to face the problem as a matter of priority in all international social policies.

7.3 Objects of this Case Study

The actions necessary to face this transcendental problem extend to basically two fields: legal and educational. In other words, acting directly upon such inhuman acts and, of course, seeking out all means to prevent them. We should not let our guard down in any country or region, as developed as it may be. The problem is often veiled. As Teelken and Deem (2013) have demonstrated in a recent and interesting article, built on a study based on qualitative interviews carried out in the Netherlands, Sweden and the United Kingdom, that is to say, highly developed countries, despite the deployment of programs aimed towards stimulating equal opportunities and reducing regimes of inequality, there persist subtle forms of discrimination which affect contexts, such as higher education, where they should have already been completely eradicated. These results agree with those presented by Silander, Haake and Lindberg (2013), and focused on this educational level in Sweden.

The insertion of gender equity in the discipline takes an urgent and primary role. In this respect, Velasco (2007) maintained that education is a vital tool for women to overcome the subordinate position to which they have been socially and historically subjected. The present article pays particular attention to the understanding of the new concepts and meanings that arise out of some discourses on equity. These can contribute greatly to practical applications that must be applied in educational fields. Discourses, due to their dialectical nature, give way to spaces of reflection, which in turn contribute to the formation and transformation of educational practices. This process, just like in the construction and deconstruction of gender, requires a construction and reconfiguration of thinking.

The formative processes of education for equality should take into account the importance of the impact that communication media represent in the socialization of population. Thus, the presence of sexist attitudes in media products should be contemplated in educational centers within the pedagogical actions that face the issue. The new technologies and communication means should be present in school curricula, and a critical analysis of its products should take place (Gómez Galán, 2007 and 2011). Undoubtedly, the influence they possess allow us to construct tendencies and social habits, at the service of enormous economic and power interests in which there is no preoccupation for the existence of discrimination against women in today's world. And all of this can be applied, in the processes of digital convergence in which we find ourselves, to all the phenomena of social networking and its decisive impact in today's world.

There is a consensus that it is necessary to continue immersing ourselves in educational policies of equality in the Western world. Figures like Mosconi (2014) and Rebollo, García Pérez, Piedra and Vega (2011) defend this point. It is not a problem concentrated in the Third World and in developing countries, as we can see, but in the whole world in general. Similarly, the economic development and globalization, at a world level, favor equality between men and women and reduce feminine subjugation. This has been shown by Potrafke and Ursprung (2012) in an interesting study that empirically measured the influence of globalization in social institutions of almost one hundred developing countries in intervals of ten years since 1970. Economic and social globalization exerts a decisive influence on social institutions in favor of equality.

Throughout the fieldwork detailed in this article, focused on the developing world and specifically in Puerto Rico, we will examine the discourses of 10 heterosexual couples who were exposed to the ideas of the feminine movement, whose experiences were broadly investigated in a phenomenological, qualitative investigation. Here, we are specifically focused on: (a) the meanings of the phenomenon of equity after receiving influences from a movement of ideological transformation directed towards equality; (b) the understanding of practicing equity and (c) how they are carried out or adopted in their respective couple interactions. Our main objective is knowing the significance of equity offered by these couples, based on the experience of practicing ideas of gender equity, for a mainly educational purpose.

There are some antecedents of international studies centered precisely on discursive practices with which to create spaces for transformation towards gender equality. An example of this is the recent contribution by DeJaeghere and Wiger (2013) carried out in Bangladesh. Focused on an educational context, they defend the importance of non-government organizations as agents of direct change in schools in the generation of discourses of equality between men and women.

The discourses reflect the mental "realities" in whose construction language plays a main role (Perakyla, 2005). Additionally, the discourses have a critical dimension, in which ideological effects are conceived as producers and reproducers of power relationships in society (Fairclough, 2005). In this context, a look at the essence of the discourses held by the participants of the investigation discussed in this article remits us to a set of attributes or values compatible with the meanings attributed to equity as a general topic. That is to say, the essence of the discourses lies in universal values contextualized in reference models that question the dominant power of some groups over others. For example, the discourses of the participants essentially evoked the following values, which are also in line with those that promote equity in all of its diverse manifestations: justice, access to equal opportunities, reciprocity, a feeling of independence (individuality), a feeling of common wellbeing (mutuality), responsibility and respect.

7.4 Research Methods

Our starting point is a larger research carried out under the title "Experiences, practices and meanings attributed to living together in a context of gender equity among a group of heterosexual Puerto Rican couples: feminist considerations for couple counseling" and other previous studies (Padilla & Gómez-Galán, 2014).

The couples that participated in the study were selected in Puerto Rico according to an availability sampling (Hernández Sampieri, Fernández Collado, Baptista Lucio, 2010) for which they needed to fulfill the following prerequisites: have lived together for a continuous period of at least 5 years with the current partner, time that represents firmness in the configurations and practices of the relationship; self-denomination as a practitioner of equitable ideas on gender roles; and availability to participate as a voluntary couple.

Data was collected through in-depth, semi-structured, phenomenological interviews with each member of the relationship separately. The interviews were transcribed verbatim, and they constituted the tool and central instrument for the following discourse analysis. For that reason, each one was read carefully and in detail using the technique of triphasic discourse analysis. This analysis consists of three basic activities: identifying discourses in text, identifying the effects of the text (implicit and explicit intentions) and identifying the contexts with relation to the meanings attributed to equity by the group of participants (Berríos and Lucca, 2003). The texts were selected to allow problematizing an issue as laid out by Michel Foucault (Berríos and Lucca, 2013).

7.5 Results

The meanings attributed to equity by the group of informants that participated in this study were grouped according to the implicit and explicit effects of the text. The meanings of equity that emerged through the discourse of the group of participants were classified in four main dimensions: [1] from absolute equality to fair opportunity, [2] from the union of two to a fair division, [3] from power to individual responsibility and [4] from constructions and deconstructions of equity towards new meanings. The construction and deconstruction of these meanings makes up the analysis presented in this article.

7.5.1 From Absolute Equality to Fair Opportunity. Not Doing Things the Same, but Having the Same Opportunities

The first discursive dimension revealed that the meaning of equity is not framed in absolute equality as recognized by the feminist movement of the 1960s. Equity did not mean doing things the same, but having the same opportunities. For these couples, it was more important that the opportunities to exercise domestic chores be equal than the execution of the chores itself. The importance of dividing domestic chores in equal-parts participation was not significant for this group. On this topic, the couples pointed out: *"There are no issues between us if one of us wants to rest while the other one does*

something at home. She doesn't bother me if I want to watch TV and I say:
look, I'm tired and I want to lie down and watch TV, I lie down and that's it.
It's the same with her, lie down and be happy." |"It's not always about doing
chores at the same time. It's OK for one of us to rest." |"It's not about I sweep
and you mop, I wash the clothes and you fold them. It's about your partner
not becoming a handicap or an unmanageable force . . ."

Within this discursive dimension, the meaning of the concept of justice
stands out. Rabin (1998) pointed out that the perception of justice in marriage
is very important because it is closely related to marital satisfaction between
couples that wish to achieve equity. The author adds that these couples
have the need to feel that the marriage is fair for them. Van Willingen and
Drentea (2001) found that the linking of equity with a sense of justice,
decision-making and domestic chores brings greater stability between the
members of an equitable marriage. In line with the arguments from the
literature for participating couples in this investigation, "fair" meant that
the division of chores was properly proportioned, which didn't necessarily
imply dividing domestic chores by half. In their respective discourses, the
couples pointed out: *"equity is for both parties to have the same rights."*
"The right to give your partner what is fair for you". Equity . . . *"has to do*
with justice." |"Right to a fair and reasonable environment in order to feel
emotionally better." "What's not fair for one party can't be OK with the
other one; you must agree to be equal and fair" |"Justice is the cousin of
equality".

These discourses also reveal an appreciative aspect to the contribution
of each party more than to the chores carried out. Their discourses reflect
an affective dimension that recognizes and values the input or contribution
of each member of the couple, more than the chore itself. Rosenbluth,
Steil and Whitcomb (1998) already pointed out that the definition of equity
should be made up of non-observable conducts like attitudes, affectivity
and interpersonal processes between partners. The authors added that the
concept of equity Is a multifaceted construct with behavioral, cognitive
and affective dimensions because it captures the psychological complexity of modern relations in marriages oriented towards the ideal of equity.
In this manner, couples appreciated the affective value in each member's
contribution towards domestic chores in concepts underlain by emotional
reciprocity. *"Equity is support in equal parts, a recognition of affect" |"Equity*
doesn't have to be in the sense of equal chore-for-chore division, but in
me feeling that what I do is also appreciated and valued by the other
person."

7.5.2 From the Union of Two to a Fair Division: Own Identity and Partner's Identity

The second discursive dimension of the meaning of equity was revealed within the vision of fairness and a recognition and appreciation in equal conditions of the link of the partner's identity and one's own identity. The group indicated that, in order to understand and achieve equity, a balance needed to be reached between one's own interests (individuality) and the interests and need of the partner (mutuality). Castro (2004) maintained that couples moving towards equity are practicing new lifestyles as partners as a result of the changes produced in society regarding new visions of gender. As opposed to the traditional discourse, the participating couples constructed a new understanding in which, in a union of two, two separate entities remained. Let us see the discursive expressions of the participants regarding this topic from very specific, and in some cases metaphorical, standpoints: *"It's about having a vision of relationship, and even when you take care of personal needs, it's a relationship to achieve a compromise between both parties"* |*"Contributing to achieve her goal and having that goal become your own as well"* |*"Contributing to achieve each other's goal and having that goal become the partner's goal"* |*"Your own, individual wellbeing becomes common wellbeing"* |*"Recognition of the capacities and weaknesses of each party, and even when you take care of personal needs, it's a relationship to achieve a compromise between both parties".*

This recognition meant that equity in a relationship requires the participation of two entities for a common good, without either one losing the right to his or her own identity. The discourse oriented towards the link between couples was interpreted as a form of social support in the relationship. In line with the experience of the participants in this investigation, Van Willingen and Drentea (2001) pointed out that marital relationships oriented towards equity are associated to perceptions of social support, because the marriage itself has been a relation of social support. Additionally, the participating couples referenced a mutual compromise with regard to fulfilling the goals, needs and wishes of their respective partners, attributing great importance to professional development. The investigations of Knudson Martin and Rankin Mahoney (1999) and Willingen and Dentrea (2001) are in line with these findings; they confirmed the importance of attributing compromise to developing the personal goals of each party in an equitable relationship.

The participating couples also manifested a need for individual space, where they could carry out actions related to their personal interests. In this respect, Castro (2004) suggested that couples moving towards equity wish to move forward with their personal goals and motivations. The discourse of individual space between couples was equivalent to the claim for personal independence. The couples pointed out, in a discourse committed to mutuality with emphasis on the nouns "freedom", "space" and "otherness": *"Equity is ... depending on oneself." "Offering space when she wants and needs it".* |*"Equity is space." "Recognizing the partner's time and space." "Freedom of movement: go where you can."* |*"Equity is having space to grow." "To be independent, achieve goals without limiting one's partner."*

Significantly, the discourse of the group of women regarding individual space stands out, where equity constituted the plain manifestation of their own identity without any disposition to give up their values, beliefs or personality traits to the dominant figure of a man. Castro (2004) found that women in transition seek a relationship based on reciprocity or equity. The author added that these women generally possess a high self-esteem, and constantly evaluate the best way to unfold their lives, the achievements and projects achieved as well as the belief in their values, personal abilities and rights. In this investigation, the women participants also pointed out that their rights were non-negotiable. This discourse of individual identity strengthened the constructions of identities referring to individuality and mutuality, at the same time that practices of equity imply arrangements and accommodations for individual wellbeing, which results in the mutual wellbeing of the relationship. The following are fragments of the discursive texts of the group of participating women in which the admission and recognition of the "other" does not undermine the satisfaction of either party: *"Equity is ... achieving goals without limiting the other." "Not being an impediment for the partner's achievements."* |*"Both paddling on the same side"*

That last metaphorical expression merits building upon: "Both paddling on the same side." This image strongly conveys equitable cohabitation, according to the discourses proposed by these couples, because it suggests conserving each partner's own (individual) identity when it points out that each one paddles at his or her own side of the boat, taking care of his or her own oar. At the same time, it recognizes the link between both parties because, when they both paddle on the same side, they are both committed to rowing in the same direction.

These discourses broke apart from the traditional ideology where the union of two people becomes one entity or unity. In this new discourse attributed to equitable cohabitation, each one is still his or herself, with a plain manifestation of his or her own identity. This discourse indicated that in equitable cohabitation, neither party stops being what he or she is in order to become what the partner wants of him or her.

7.5.3 From Power to Individual Responsibility: A Discourse of Men Willing to Cohabit Equitably

Power for the group men participants implied a willingness to incorporate patterns and constructions of equity in their own lifestyles. This implied that they were willing to break away from the traditional role as dominant entities. In this manner, they showed a greater openness and availability to free themselves from the oppressive structures of the traditional social construction of gender. In this respect, the men pointed out: *"neither of the two parties feels uncomfortable in the role they have to carry out"* |Equity is . . . *"For my partner to have just as much power as I do in the relationship"* |*"Neither to dominate the other, nor for the other to feel dominated"* |*"Effort that one makes to reach the point that necessarily occurs in equal parts or doesn't occur"* |*"For each one to be on the other's side, not leaning on the other."*

In turn, the men attributed great importance to their own responsibility, whereby they deliberately decided to abolish their traditional role as the dominant part. Equity as own responsibility meant that this group of men was capable of assuming actions and chores without being unfair towards their partners. Their ability to assume responsibility towards themselves was as a manifestation of their own identity. That is because, assuming their responsibility, they have the opportunity to manifest their beliefs on gender equality through their acts. On this topic, the men in the participant group assume behaviors considered as equitable, even in opposition to the traditional view of the "macho": *"Equity is something that is one's own responsibility, it is not about wanting to deny one's own existence"* |*"I very much believe in individuality. That is to say, you are you, and you have the potential to do things right"* |*"The message that we must send to 'macho' men is this: assume your responsibility because it is yours and nobody else's"*

These discourses support the statements by Castro (2004) which support that men in transition mostly belong to new generations that have associated themselves with reference models that differ from the traditional viewpoint. According to the author, they have been better prepared to engage in equal relationships that respect the characteristics, needs, wishes and projects of both parts. These men have perceived and accepted to change the disturbing effects of gender stereotype in a relationship, and they have been willing to redistribute power, which means accepting that both parties should enjoy an equal power quota. For this reason, one of the participants distances himself and speaks about the *"message that we must send to 'macho' men."*

7.5.4 From Constructions and Deconstructions of Equity Towards New Meanings

The discourses offered by the participating couples revealed new meanings of equity, to which a dynamic, reflexive and learning character was attributed. These meanings were contextualized as new understandings oriented towards the social construction of equity. This dynamic quality stands out in several facets: diversity, dialogue and reasonableness, whose elements characterize the discourse: *"Equity doesn't mean that we have to do everything in equal parts. That's not true, that's not literal. It's a matter of you listening more than talking. And not ceasing to talk, it's not about closing yourself off and not expressing yourself. I think it's something dynamic."|"Contributing so that the other one's burden is not so heavy, and accepting that some days the priority is not yours. Accepting that some days the priority is not yours, and that you are not always right" |"You learned, and you still learn step-by-step. Equity is a learning process, continuous. It's about trying to understand spaces, and trying to maintain a balance within those spaces."*

All couples admitted that, even when they perceived themselves as having achieved equitable practices in their respective relationships, there was space to improve them because living together equitably has required a great investment of time and effort, and a constant evaluation and monitoring. This discourse presented new connotations to the meaning of equity between couples because it suggested that it is constructed and reconfigured according to the partner's experiences.

7.6 Conclusions

The challenge of equality is still one of the main objectives of society. Inequality between men and women is still a first-level problem, even in more developed countries. It is not a problem of the past or of specific geographical spaces. It belongs to the present, and it includes all nations. Additionally, the social dimensions where it is present are very widespread, and beyond being limited to the economic, professional or work field, it generates phenomena as degrading for humankind as gender violence, with all that it implies at a world level. Without mentioning, of course, the level reached by discrimination against women in particular cultural areas, which leads to actions as cruel and vile as feminine infanticide, ablations, death penalties for reasons of gender, etcetera.

One of the most important tools our society must use to face these problems is, without a doubt, education. Education and legal action constitute the two fundamental pillars which should allow us to move forward. Equity must form a part of all educational processes, and it must be present in all international social policies. It is vital to keep one's guard up. Even in more developed countries, things like labor discrimination, sexist acts or even gender violence are much more widespread than one would suppose. It is a universal phenomenon. It is necessary, therefore, to promote education for equality in order to prevent it and, in the shortest possible time span, make it disappear.

Our case study is framed on the knowledge necessary for educational actions. The information that we can obtain regarding living together as a couple in the developed world, and the meaning attributed to equity, is fundamental. With the objective of broadening our vision for its integration in educational contexts, we have focused on the discourses of those couples that participated in this study because they represented various ideological constructions and deconstructions based on a new reflexive and dynamic understanding of equity from perspectives of fairness and otherness regarding diversity or difference in a relationship. Their discourses developed according to the interpretative reality of this particular group. Their notions of the concept originated in processes built from their own interactions with their partners in a historical and cultural context like that of the Caribbean, which resulted from social ideologies of feminism as an organized movement. As a result, their discourses were the product of positioning partners as subject-participants of their own reality.

Each couple attributed meaning according to experiences lived during a process of new ideological configuration, even when they all held affirmative beliefs and considered themselves as practitioners of equity. New constructions of equity stood out in this discourse, such as: (1) a compromise to move in the same direction and have the same opportunities, more than doing things the same; (2) recognizing and valuing the relationship in equal conditions more than individual needs, in order to create a fair balance between individuality (I) and mutuality (partner); (3) distributing power in the relationship that has been socially conceded to men in a willingness to yield their comfort zone and share it with women; and (4) monitoring and evaluating the relationship, viewing it as dynamic and as entailing learning in order to continue constructing and deconstructing equity.

These discourses were positioned according to rights, justice and the construction of new identities, or a consciousness of one's own subjectivity. This discourse of equality required not just construction, but also deconstruction of both identities and actions where a sense of liberty is manifested, because practices of equity are liberating in that they break away from the oppression of some human beings over others. It is about processes where freedom is constructed and reconstructed in the face of oppression: if the shoe is tightened, we loosen it so as not to hinder the way.

8

Climate Change Education: A Theoretical Model

José Gómez Galán

Metropolitan University, PR, United States,
and University of Extremadura, Spain

8.1 Introduction

Environmental issues have become today one of the major challenges facing society. Although multiple dimensions and approaches are needed to address this urgent challenge, certainly education should have a special role. Among all areas of human and social knowledge there is one on which we can work not for today but for tomorrow, because the actions and behavior of future generations are being built together with it. Indeed, this is the foremost objective of education, which should be the main driver for improving our world, the present and the future. And it is absolutely essential for the preservation of the environment, respect for nature and the fight against climate change. As defended Freire (1984, 2006 and 2014) education involves a critical economic understanding of social reality, politics as well as the implication of literate environments. In this sense, environmental problems, so powerful in our society, cannot be left out. Unfortunately, in many social contexts education is still considered exclusively as an instructive process focused on acquiring knowledge of cognitive nature by each learner, assuming that it will be critical to their maturation and future integration into society. However, for the creation of a responsible citizenship and shaped to the needs of the XXI century world it is essential to address comprehensive training processes in which not only the cognitive dimension is taken into account, but also the procedural and, above all, attitudinal dimension.

In this sense, preservation of nature must not only rely on the knowledge of the situation that our planet is globally facing, but it is up to the whole of society to act and participate decisively through actions and behaviors in such a necessary goal. It includes not only requiring the most powerful companies and industries or governments of nations that among their plans for economic and social development they provide environmental issues, as if they only had the ultimate solution, but also knowing deeply that the present situation of nature is not a problem disconnected from others affecting all human beings, but a cause of the evolution of our society. It is the price that the Earth is paying for our evolutionary development, and it is therefore quite wrong to consider it as a watertight compartment separated from all the dimensions that affect humans. Also, facing this problem effectively can only be achieved when we all change our attitudes, without distinction, not only of those who, for one reason or another, have a greater impact on the environment.

From an educational perspective the problem has been considered for the last few decades and the results are training processes grouped in the so-called *Environmental Education* (EE). And based on it, it becomes urgent to develop genuine environmental values that will change attitudes and behaviors leading to sensitization and awareness of such serious problems, which must be appropriately integrated in educational contexts. But are these really its powers and functions? Is this the main objective of the EE today? Our starting point is previous researches (Gómez Galán, 2008 and 2015a).

Indeed, since its start Environmental Education was seen from an interdisciplinary perspective, participating in all areas of knowledge and with the aim of reaching all possible formative dimensions (including, of course, ethical values). We consider, of course, that it cannot be otherwise. This is the only means for genuine development. At the Tbisili Conference (UNESCO, 1978) it was specified that *natural* and *created environments* are the result of an interaction of multiple aspects (biological, physical, social, economic and cultural). Together they must be taken into account in the process of training for the preservation of nature and respect for the environment.

It is certainly its main goal, and only like this will it become possible to establish the cognitive, procedural and attitudinal dimensions necessary for a complete education. It is also appreciated that, at that time (let's remember that we are in the seventies), looking ahead at ecological, political, economic interdependence in our world, as well as a supranational consideration of the issue, was established as one of the essential objectives of the EE. Precisely, the efforts to improve the environment could affect dialogue among countries and the creation of a new international order.

In this context, the teaching of human natural bases and their direct relationships with the environment must be implemented alongside social, cultural, and once again, ethical dimensions. The EE must be taken into account from all aspects of human knowledge, in a globalized and globalizing way, and contribute to a better understanding of our world. It is impossible to understand our society, our development, our lifestyle, if we don´t consider, in short, what we are and what our origins are. All evolution is impossible if we are unable to maintain the space that holds our lives. It must be understood that nature is not in any way, as it is sometimes considered by many primary school students who live in urban environments, and quite a few adults, a separate area distinct from the city, almost independent, and usually a place for recreation, but we are talking, above all, about the essence of our existence. And this, of course, has first-level ethical connotations, embracing, directly or indirectly, what may fall into the field of *Values Education* (Gómez Galán 2008). Other researchers such as Smyth (1996), Bianchini (2008), Ferkany and Whyte (2012), Pe'er, Yavetz and Goldman (2013), and Kronlid and Öhman (2013), also participate in this view.

Therefore we assume that it is essential to view environmental education from an interdisciplinary perspective. This was proposed in Tbisili Conference and we should not lose sight of this horizon. However, today's situation is not like that. In recent times, although it was mainly from the nineties, we can see a trend in which the EE is losing that initial comprehensiveness in their educational approach and is moving to specific fields, especially in Science teaching. Why is this phenomenon taking place? There would be different explanations for this, but we'll focus especially on the two approaches proposed by Sauvé (1999): on the one hand, environmental education is not rewarded today as it is associated with elements of social and educational criticism, questioning common ideas, and tends to be offered within a specific framework, losing its authentic tangentiality and merely focusing on environmental aspects; and secondly, the fact that lately different proposals of comprehensive education are being carried out, including all essential aspects of human development, among which it could have a place, but presented as a portion of these, and therefore incurring in a sharp reductionism.

We can mention as the most important examples, almost all of them born in the nineties but with large current development, of these new integrated approaches, (1) *Education for Sustainable Development*, whose general foundations have been widely exposed, among others, by Bonnett (1999), Halfacree and Ellison (2001), McKeown, Hopkins, Rizi, and Chrystalbridge (2002), Apel (2007), Barth, Godemann, Rieckmann & Stoltenberg (2007), Venkataraman (2009), Zenelaj (2013), Scott (2015), Thakran (2015), and Berryman & Sauvé (2016). Derivatives thereof are (2) *Education for a Sustainable Future*, widely defined by UNESCO (1997) and developed and analyzed, among others, by Hernandez & Mayur (2000), Fien (2003), Calder & Clugston (2005), Rowe (2007), Kirchhoff (2010), Pavlova (2015)

and Bostad & Fisher (2016), or (3) *Education for Sustainability*, described by
Tickell (1997), Sterling (2001), Jämsä (2006) or Sterling and Huckle (2014).
The origin of these expressions, as advocated by González Gaudiano (2007),
took place in 1992 when the term "environmental education" was abolished in
the wording of Chapter 36, Agenda 21, in an attempt to give greater importance
to the concept of education as a means to prevent the general deterioration
of the world. But the results were not as good as expected, and the approach
offered had a lot to do with that.

In these new proposals for comprehensive education all the issues involving sustainability try to be globally faced, from multiple social and economic
dimensions, and precisely here lies the main problem as the care for the
environment and respect for nature in a context of economic growth is
considered as the engine of social development and of our civilization, based
on the idea that all progress is based on it. It is accepted as a fact that
the increasing exploitation of our world will continue (without taking into
account that the Earth has limited resources) and, to be truly effective, it
must be sustainable above all, thinking about future generations. In short,
it is an anthropocentric approach. Several authors have, precisely, analyzed
the presence of this perspective in these most recent educational constructs,
which seek to overcome the original postulates of Environmental Education.
In addition to our case (Gómez Galán 2005, 2008 and 2010) there are examples
such as Kopnina (2013 and 2014), Clement and Caravita (2014) and Imran
Alam and Beaumont (2014).

It is true that different educators use these terms for proposals that we
can't definitely understand as anthropocentric, but it is often due to lack
of information about terms and contextualization. As Cocks and Simpson
(2015) have recently demonstrated there is a long way between environmental
philosophers and educators who put it into practice. However, despite the fact
that once Bonnett (2002) argued that the problem is beyond anthropocentric
or ecocentric concepts – we support this standpoint, in line with Tortolero
Chavez (2004), biocentric and zoocentric-, the truth is that in the case of global
educational policies (such as UNESCO, the European Union, departments
or ministries of national or regional education, etc.) the approach is very
important, because it determines general guidelines and programs, above,
obviously, the ethical implications. Thus, in the current framework presented,
Environmental Education no longer has its original globalizing meaning
and appears, in many cases, within a separate compartment that seeks to
avoid all kinds of connotations critical and focuses, almost exclusively,
in conservationist matters basically, in order to teach different actions for

the preservation and cleanliness of the environment (recycling, responsible consumption, waste treatment, energy issues, etc.) but not a direct approach to the causes of its destruction (and it leads it, especially today, to position itself in the field of Science education).

In this context, in which any criticism of the progress of our civilization (produced greatly at the expense of environmental degradation) wants to be avoided, essentially for economic purposes, some fundamental scientific or epistemological aspects are ignored and, what's more essential, ethical, particularly in relation to the exploitation of some people over others, or of man against nature and other living things (going beyond the exclusively anthropocentric vision which is dominating today).

Thus, we need to reformulate a proposal of global Environmental Education, a new philosophy that allows us addressing environmental issues from an educational perspective and to its fullest extent, so that we can consider objectives focused on knowledge, attitudes, skills, participation and, above all, the awareness of the destruction of nature and so returning to its original interdisciplinary conception. In other words, resuming its link with an education in values, increasingly required by the delicate situation in which we find ourselves, and also fully develop their multiple connections with many other scientific and philosophical disciplines that have a direct or indirect need for a productive dialogue.

8.2 Main Objective: Redefining Environmental Education (EE), Education in Environmental Values (EEV) and Climate Change Education (CCE)

We must insist, therefore, in maintaining the interdisciplinary nature of environmental education, regardless of their presence in other proposals for global education. Moreover, in our case, the *Environmental Education* (EE) World be considered as one of those proposals that seek to deal, in a general way, with all the problems we face today, and they should be reflected in a comprehensive manner in the training process. Its original tangentiality must be employed effectively and urgently, in educational processes because we start from the fact that we face a discipline with tradition, which is not the case of some of the new educational proposals. Today, we must seek their allotment rather than its expansion. We are certainly defending the need to recover the authentic holistic dimension of environmental education, especially in its relation to values education, perceiving the preservation of the environment and combating climate change and respect for nature as well as the living beings that inhabit it, as key objectives of our society, which must involve the attitudes and behaviors needed by citizens to their achievement. We need to recover the term *Environmental Values Education* (EEV) (Gómez Galán 2008), appeared in the seventies and eighties of the last century (Kauchak, Krall and Hemsath, 1978; Knapp, 1983; Caduto, 1983 1985) but little used since then (Scott and Oulton, 1998; Hartsell, 2006; Gómez Galán, 2008 and 2010), and it would involve a direct reference to the ethical dimension of it, to the specific features of its application in educational backgrounds.

Of course, the effective implementation of its basic objectives, or a new reformulation in the terms we are defending, involves the development of a new teaching methodology. And, in this sense, teacher training is crucial. Changes in the nineties, in connection with the integration of EE within the overall proposals for sustainability, led to its positioning, as we have underlined, in the area of Science Teaching or Science Education, performing scientific development, in a very general way, in the various departments of Science education at schools or faculties of education, or even those belonging to the Schools and Faculties of Science, directly connected with ecology and environmental sciences but without any link with pedagogy, education and, in general, social sciences and humanities. A situation that some authors, like Tytler (2011) or Pedretti (2014) looked at from different approaches but, above all, warns us of the complexity of this problem and its situation today.

Whereas there is no doubt, of course, that their epistemological bases place their knowledge structure in the field of natural sciences (and their training is essential in teaching professionals), we run the risk that, if this is done in an exclusive manner, its direct connections with other sciences (mainly social and human) may disappear and greatly minimize its application dimensions, especially all those associated with values and ethics. It is not only a scientific or knowledge issue. This reductionism in the concept of EE is not possible. Addressing now the serious environmental issues is essentially a matter of values (Gómez Galán 2008). A reorientation in the essential characteristics of EE and teachers training is essential. There is an urgent need to ensure an awareness of citizenship, which leads to an empathy towards nature and the living beings, as well as all of the human populations in greater danger of being subjected to the consequences of environmental degradation and climate change, and that will only be possible by carrying out a deepening on all major issues addressed by environmental ethics.

In this context, the education "is an essential element of the global response to climate change [. . .] helps young people understand and address the impact of global warming, encourages changes in their attitudes and behaviour and helps them adapt to climate change-related trends" (UNESCO, 2010). Thus, *Climate Change Education* (CCE) should be part of EE/EEV. It is a new term used in diverse sense (Lassoe, Schnack, Breiting, Rolls, Feinstein & Goh, 2009; UNESCO, 2010; Baek, 2012; Blum, Nazir, Breiting, Goh & Pedretti, 2013; Koya, 2013; Beatty, Feder & Storksdieck, 2014; Fernandez, Thi & Shaw, 2014; Hung, 2014; Boakye, 2015; Mochizuki & Bryan, 2015; Trajber & Mochizuki, 2015). Blum *et al* (2013) argues "that such debates about how EE, ESD and CCE are conceptualised remain highly relevant [...] also to wider international discussions regarding both the current and potential relationships between conceptual understanding, policy and practice". Overcoming reductionism in the concept of Climate Change Education is needed; it must be integrated into a broader notion of environmental education. Of course, it must be viewed from an ethical perspective.

Focusing on an *Education in Environmental Values* we will be offering a new scientific and didactic perspective that takes into account all of the elements that we are presenting, starting from a philosophy of comprehensive education (as set out at Tbilisi Conference). And especially at the present time, because we cannot forget that we are in a process of convergence in education worldwide resulting from globalization. All this implies a reorientation of the EE, recovering its initial objectives and adapting to a new framework for the twenty-first century, in which it is unquestionable that, in addition to the degradation of the environment, we are in a period of climate change of anthropogenic origins whose consequences are difficult to predict, but which will undoubtedly have a decisive impact on our planet and, especially in those countries with fewer resources (Bellard, Bertelsmeier, Leadley, Thuiller and Courchamp, 2012; Fankhauser, 2013). We need a new model of EE that avoids reductionism and drinks from its original sources but adapted to new needs, keeping a firm hand of the rudder of ethical values. This requires understanding the whys of the many current proposals and what are the reasons why those with a strong anthropocentric approach have a greater presence. From that knowledge it will be possible to carry out this transformation and define its characteristics to meet the new challenges of our society.

8.3 New Theoretical Model: Multidimensional EE in Ethical Context

Therefore, the dialogue between environmental education and ethics is unavoidable. This gives rise to many questions that need to be answered: how do you take into account the ethical dimension today in EE? How does it appear, from a pedagogical perspective, in current curricula? Are environmental issues, in all their scope, present with all their ethical implications? What the connections today between EE and Environmental Ethics? All these questions tell us about the intrinsic value of integrating ethics into a multidimensional Environmental Education. Alternatively, the intrinsic values that were already present in the original conception, as noted above, should be recovered. As we are talking about a transformation, the merger of these elements will lead us to a new context of environmental values education that may address educational needs and challenges and so responding to current social demands.

We are encountering again, in the middle and the second decade of the century, the need to integrate ethics and values in educational and pedagogical processes of EE, as well as a firm commitment with its multidimensional perspective, which has been lost in recent decades. Construction should be from multiple disciplines, as the epistemological foundations of science education alongside the main concepts of ecology, environmental science, sociology, political science, history, economics, geography, etc. are necessary. And we should not forget philosophy, as an essential foundation of ethics. If we do not understand EE processes as processes primarily for raising awareness of citizens to the many problems of our actions on the environment, or we do not assume the key causes of climate change, the complex social problems it generates -and will generate-, and the status of our relationships with other living beings who share their existence with us on this planet, we will be unable to make any progress or improvement. Regardless of the intrinsic damage and heartrending pain that we are producing on so many sentient beings - our own species included-, unless we change our attitude towards the whole of the biosphere, we are set to suffer the harshest consequences (Gómez Galán, 2005 and 2010). No doubt, we are also talking about multidimensional ethical issues that must be integrated into the educational process.

Due to this complexity, which is inherent to this set of problems, we argue that it is the key to explain the existence of different trends of environmental education, involving the confusion, distortion and inadequate practical application in educational contexts (with a predominance of some of them in time, as we have noted,). The reason is that we believe that its origin lies in the different philosophical and scientific trends that have tried to explain the relationships between human beings and the whole of nature, the behavior and attitude that we must have, the moral problems arising from our actions and ultimately our behavior *in* and *with* the environment. This is, of course, ethics (the branch of philosophy that studies and analyzes moral actions, that is, the set of customs and rules that govern our behavior).

And when we look at it in the context of the environment it can be defined as Ethics of Nature, Ecological Ethics or Environmental Ethics, term most widely used today, in the sense used by authors such as Sosa (1990), Gomez Heras (1997), Minteer (2012), Hourdequin (2015) or Pojman, Pojman & McShane (2015). The term we consider most appropriate (although in recent years it has undergone a process of reductionism practically identified with biomedicine), Bioethics (ethics of life, to which we belong inseparably, so we should avoid talking about independent ethics related to the environment: we are also the environment, we must never forget), as used in this sense by Potter (1971) - who originally proposed the term-, Robles (2000), Sarmiento (2001 and 2013) Macpherson (2013), Escobar & Ovalle (2015) or Thompson (2015). In our case we prefer, in the analysis concerned, the term Environmental Ethics, as we aim to establish its relationships with Environmental Education, from which it cannot be dissociated.

Thus, what are the major mainstreams of environmental ethics, from which, as we deem appropriate, the various proposals for Environmental Education emerge? To answer this in brief, and simplify the analysis, we will rely on the classification established, although for a different purpose, by Chavez Tortolero (2004). Adapted to our goals and views, we agree with this author that we can identify four lines or general trends whose general characteristics, schematically presented, are as follows:

A. *Anthropocentric* (human-centered perspective): their welfare, happiness, security, etc. Very close to the moral right of the Western tradition. This trend holds the controversy of the specific profile of what mankind is and our relationship with the rest of nature and the beings that inhabit it.

B. *Zoocentric* or *Animalist* (focused on animal rights, regardless of the specific human rights): the main idea of this trend is that animals should partake of basic rights: the right to life, not to be tortured and live in their environment.

C. *Biocentric* (life-centered perspective): in this trend there is a moral egalitarianism among all living beings, including humans.

D. *Ecocentric* (focused on ecosystems and biodiversity): this trend is concerned with the preservation of species and biodiversity. In other terms, it is interested in maintaining the integrity of biotic communities and good balance of ecosystems. Animals and plants would not be the only ones with moral consideration, but also water, rocks, air, etc.

We argue that all these currents in the field of environmental ethics are those that have ended up, in different ways, in different proposals for

Environmental Education. Among all of them the anthropocentric tendency, for many reasons, has been most successful, and fully explains the fact that the different educational policies on the environment adhere to it. Of course, it is not just because it lies on philosophical pillars (and even religious) of long tradition in the West, but also because it is more suitable for multiple social, political interests and, above all, economic, as we have had occasion to see. Removing animals, plants and, in general, ecosystems from any moral consideration is, no doubt, extremely profitable for the powerful. In that context it is not productive for the population to be critical of their companions, in view of many of the most dreadful actions of human beings, in this common home, which is our planet (Gómez Galán 2010).

A paradigm shift is required in the basic concepts of Environmental Education, too anchored in the anthropocentric current. If there is no direct participation of other trends, the future of humanity itself will be at stake (it is an indivisible part of environment in which we live). We must take full advantage and the new insights offered by each of them, minimizing its drawbacks (e.g. trying to reach goals defined by some unrealizable utopian ideas), thus enabling education strategies leading to reflection and action in order to show the situation as it is, without the presence of as many demagogic elements are shown in many training processes. Beyond the Environmental Education is Environmental Ethics, and certainly this greatly affects the characteristics thereof. We are talking about an ethical issue -above all its many manifestations-, ultimately linked to values, and without the presence of this dimension we won't fully be able to speak of progress and improvement in the face of the serious situation in which we find ourselves.

Among all the trends referred the anthropocentric movement is the one which provides lower educational benefits (and therefore for the evolution of humanity) in its application, as it is mainly governed by principles of economic development over ethical values. As noted above, it is offering extremely reductionist models of EE, often highlighting only very specific actions and practices (recycling, energy consumption, etc.). Of course, in these situations, the aim is not being critical of the predation that the human being, far beyond their needs, causes on the environment and other living beings, and the great inequalities that it causes among human beings themselves in different countries. Whereas certain nations (what we would call the First World) deplete the Earth's resources and generate highly polluting and harmful waste to our home, others barely have the minimum to survive. It is a scandal that in our society some countries generate unlimited waste and simultaneously lots of children die of hunger. There is no ethical justification for it and it warns us of how mu ch we still have to progress in consciousness, awareness and values.

However, while this is extremely serious, a reductionist and anthropocentric approach towards environmental education can also lead to extremism, even sheltered under the umbrella of an ecological or environmental conception, contribute to further complicating the situation and move away from a much needed global consensus. An example of what we are discussing is that only a very limited and too anthropocentric EE, or the complete absence of it, has undoubtedly allowed the development of a phenomenon like creationism, which is acquiring in recent years increasingly importance in the US, especially for its educational dimension concern, which has led even to the extent of discussing in some places the foundations of the theory of Darwinian evolution and the convenience of its presence or not at school in educational syllabuses (Berkman and Plutzer, 2010; Hickey, 2013).

Obviously, only a minority accept the literal interpretation of the Bible, advocating the creation of human beings as presented in Genesis. Not going further, other more contained trends, such as Intelligent Design, in dialogue with science and where the theory of evolution is advocated, even specifying that it is directed by God, has brought about a heated debate that has transcended science and philosophy specifically moving to the scene of social conceptions (Dembski, 2002; Shanks and Green, 2011; Brigandt, 2013; Sewell, 2015). This movement, as stated Brumfiel (2005), is a minority group in American campuses, but the fact remains that it has generated a high impact. In Europe, as we showed in due course (Gómez Galán 2008), the controversy is not relevant, but it does not prevent the fact that we are facing a situation where it is not easy to establish a global framework of action. In today's society it is more necessary than ever a dialogue faith-culture, which is advocated, of course, by any moderate stance. But we are describing a situation in which positions can be radicalized, leading to extremely worrying situations. Religion does not have to be confronted with science. On the contrary, every thought and prudent theological reflection, of course, would defend it. However, we should not forget that religion is very often used in the service, among others, of social, political and, of course, economic interest.

In this context the real problem obviously occurs when all goes beyond the scientific, philosophical and theological dialogue and a particular trend in the educational field is supported, being a part of the exclusive educational processes of children and teenagers. Creationism, especially when carried to an extreme of full scientific denial, is a good example of this: radicalization and ultimately anthropocentrism. As a part of the environmental education, and regardless of the theory of Darwinian evolution, this movement could see nature only as a resource available to the human being, and nothing else, a garden you have to look after and enjoy, irrespective of considering if we are the garden owners or the rights we have on it. And in so doing, and rejecting scientific knowledge, we should be aware that our garden isn't actually only ours, the whole garden is our home and our survival depends on it (Gómez Galán 2008).

Our commitment to a full EE, as mentioned above, based on the construction of an authentic education in environmental values, focuses on merging nature ethics trends previously presented. And this implies a critical analysis of the current situation in an interdisciplinary and multidimensional context, thus eliminating any radicalization of any kind or nature. Therefore, a global and inclusive approach is necessary, starting from values common to all human beings, implying an intercultural, inter-religious, moral and social dialogue. As we can see, the EE must be present beyond a particular environmental ethics, and should be much more than a set of teaching strategies focused on certain activities or practices. It must be a construct that allows reflecting on the world in which we live and how to improve it, including the whole of nature in our ethical scope of action. And this entails a profound transformation of what is now being done.

8.4 Climate Change Education in Environmental Education

We assume, therefore, the need to recover the essence of the first proposals for environmental education while, in parallel, the necessary adjustment must be carried out to emerging issues and needs in social and environmental areas, across all its strands. In this sense, for example, today we face new problems that did not exist in the origins of the EE or have no relevance today. Just think, without going any further, of the need today to educate not only for the preservation of the environment but, specifically, to tackle global warming and climate change, whose effects can be quite dramatic for our civilization. Similarly, we could talk about the new educational frameworks present across the world, which require a methodological and curricular restructuring of the discipline (Gómez Galán, 2008 and 2014a); the actual power of technology and media (especially with the advent of the Internet), as key elements in the processes of formal and informal education and in which the EE should have a significant relevance (Gómez Galán and Mateos, 2002; Ardoin, and Kelsey Clark, 2013; Gómez Galán, 2015b); the effects of globalization not only economic but also cultural, with all that that implies in creating educational models to be used in a global context (Gómez Galán, 2014b); and so on.

As a result, we are facing new challenges in the field of EE. New aims must be considered in the context of the education in environmental values we advocate, and even they would acquire a strong role. In this regard, we'd like to dwell particularly on four of them for which we believe there is a special urgency. New challenges that must be inherently contemplated in the multidisciplinary construct assumed by environmental education today:

8.4.1 Climate Change Education

At present there are many studies empirically confirming that we are in a period of abrupt climate change caused by human causes, with a substantial rise in the average temperature of the Earth (global warming). It wouldn't make sense, therefore, to mention the abundant scientific literature thereon. Moreover, we have the latest IPCC reports (Solomon et al, 2007; Pachauri and Meyer, 2014), which summarize the conclusions reached by many specialists in diverse scientific areas around the world, and analyze, from all points of view, present climate changes. The conclusion, consensual and practically as never in science, is that this climate change can have direct, and certainly catastrophic, effects on the human population and the entire biosphere.

Although it is a problem of vast echo in the media (Nosty Diaz, 2009) we cannot say the same about its rigorous presence in educational environments (Gómez Galán 2010). Obviously, we must educate to deal with climate change, a challenge that is still far from reaching educational practice. And, on top of that, there are too many scientific and ethical justifications that would force us to do so. It is essential not only to inform citizenship, but also to carry out full training in a process of sensitization and awareness (we insist that stopping it concerns not only to governments; the active participation of all of us is essential) of the enormous importance of global warming in which we find ourselves.

Besides being a catastrophe for many ecosystems, a process of unprecedented ecocide and biocide (the less due to anthropic action, with no natural disaster), which already would be exceptionally serious, it is necessary to show that man himself, as a member of the biosphere, cannot escape its consequences. And especially fragile are the least developed countries and the Third World, which do not have sufficient resources to deal with all this. Droughts, floods, mass migrations, extreme weather events, desertification, spread of diseases, etc., can be dramatic for hundreds of millions of people.

We must not forget that our civilization emerged and adapted to the climatic conditions of the Holocene, a period of more than ten thousand certainly stable years, although slight variations have been produced (such as the Medieval Warm Period or the Little Ice Age) allowed the Neolithic Revolution and the growth of our species to become what we are today. The alteration of this relative climate stability implies an adaptation for which we are not yet technologically ready. The only way to avoid its consequences would be facing the progressive global warming which took place especially after the Industrial Revolution, i.e., anthropogenic (Gómez Galán 2010). Thus, authors such as Crutzen and Stoermer (2000) even speak of a new geological era, the Anthropocene, the result of the influence of human activity on Earth.

8.4.2 Alternative Energy Sources Education

The scientific and technological education of citizenship is essential. We need to pursue science as a means to develop alternative energy sources to the current fossil fuels, primarily responsible for environmental degradation and climate change we have referred to above. Creating confidence among the population about the possibilities of science and technology will contribute to a social demand to support research funding policies on alternative energy.

Of these (as the current wind, hydro, ocean, geothermal, etc., regardless of future, for example, nuclear fusion) without a doubt the most important one has to be the sun. The sun is an inexhaustible source of energy, and using it is only a technical issue. As Kalogirou (2013) points out, the origin and evolution of humanity has been based on solar energy. And that should be our future. Of all the alternative energies it is the only one that offers greater benefits and fewer drawbacks.

Although the results obtained with solar energy have proved successful (Solangi and others, 2011) there are strong interests to underestimate its abilities and promote the maintenance of fossil fuels as the main energy source. One example is the extraordinary development that the technique of hydraulic fracturing (fracking) has had for the extraction of hydrocarbons, which has even led to lower prices of these. In the case of the United States, for example, it has been increased tenfold in certain products between 2006 and 2013 (Perkins, 2014), which does not only have negative effects on the environment, but also slows the scientific and technological efforts, not being economically cost-effective, in the search for alternative energy.

The solution to all this, no doubt is once again education. Rigorous scientific training allows citizens to be aware of the high price paid for the maintenance of an energy model like the present one, based on fossil fuels, and the need to invest in alternative energy among which, as mentioned above, solar (in different technologies) would be the most appropriate. Unfortunately, as we showed, training of students is not taking place today (Gómez Galán 2008).

Furthermore, this is the only means that can be used to counteract the strong media pressure, due to the major economic interests behind it all, against renewable energies, as analyzed Scheer (2012). But the reality is quite the opposite. The potential of alternative energy current model is immense (Armaroli and Balzani, 2010). They would not only contribute to the environmental objectives, so necessary today, but also they would offer overall picture in the social field, for example by creating an extraordinary number of jobs in the service of these technologies and the new energy paradigm. But to make this possible, many political decisions should be made thereon, which can only be achieved by increasing citizenship awareness and, with their decisional power, force to do so.

8.4.3 Animal Protection Education

The integration of education for animal protection in the new EE we propose is absolutely essential. We are talking about an education in environmental values and our planet holds a large number of sentient beings (with the ability to feel pain, fear, anxiety, etc.) which are also earthlings just as we are, children of our planet. The ethical dimension of the problem is immense.

We're not just talking about the animals that live in the Earth's ecosystems, in complete freedom, and receive the impact of our actions on the environment. We also refer to those which are at our service and help us to feed ourselves, provide us clothing, entertainment, etc. Overall, the damage inflicted on these creatures is absolutely intolerable. It is so dreadful to witness the situation in which, every day, billions of sentient beings are crammed into industrial farms, experimenting centres and laboratories, participating in public performances, and some others, that ethical essence of what we understand as humanity, of what we are as a species, is clearly called into question.

These creatures, slaves in the hands of a super-predator, are subjected to situations and acts with such a great suffering that any description with words would be absolutely impossible. Such a lack of compassion or sensitivity is difficult to understand. Especially because for this purpose, despite the fact that some people claim that all these actions are essential to maintain our standard of living, in all its dimensions (which would also be questionable), there are now alternatives. But, as usual, power and economic interests prevail over the ethical and moral values (Gómez Galán, 2005 and 2010).

This situation has been systematically analyzed by authors like Regan (1993), Singer (1995), Mosterín (1995), Bekoff and Goodall (2003), Gruen (2011) and Tester (2015). Clearly this is a major problem whose main solution lies, as in many other issues, in education. The main objective would be to create, mainly in children and young people, empathy for other animals, and provide them with a dignified life and the right not to be abused as sentient beings that are capable of suffering. Our circle of compassion must also cover the creatures who share with us their existence on our planet. As demonstrated in due course (Gómez Galán, 2005 and 2008) this should be one of the most important goals in the context of ethical and moral values in the EE model that we defend. Mistreating these beings, inhabitants and brothers like us in the biosphere, denigrates all as human beings.

8.5 Teacher Training for Multidimensional EE

One of the fundamental challenges of this new integrated environmental education presented would be, no doubt, teacher training. To effectively achieve any of the major goals that define the model, adapted to the urgent problems we face, it is crucial to prepare teachers for this, as they must be the principal agents of change. Without the pedagogical training of teachers who should carry out this work everything would be in vain.

Any education process lies on practice, that is, the development of action in operating contexts. As White (2005) demonstrated in due course, EE for teachers poses a theoretical basis to support the ultimate goal, which is practical action. The key question is: are teachers now willing to develop learning processes based on a new theoretical model of EE? All of the investigations lead to an affirmative answer.

Starting from a thorough understanding of the situation in which the group of teaching professionals concerned with ecology and the environment is found it is possible to determine what are their interests, motivations, concerns for problems, scientific and didactic training, trust or distrust of institutions, the way that they are facing environmental problems, etc. Particularly in Spain we conducted a complex study (Gómez Galán 2008) which allowed us determining, and we specify to the maximum in this work, that teachers are very interested in environmental issues and are aware of the damage we are doing to the biosphere. They are highly motivated, as well as concerned, to address this serious issue. It is also interesting to contemplate our relationship with nature in an ethical dimension, and consider as necessary the existence of values that allow a suitable behaviour with the environment and other living beings that inhabit it (responsibility, respect and solidarity).

However there are different barriers to these intentions. One of the main ones is that training is clearly insufficient, especially from a scientific perspective (in our study we found out that a significant percentage of teachers surveyed had uncertainties about the theory of evolution of Darwin). They are also highly influenced by the media and participate in various topics or misconceptions. No less important it was to see how the intense teaching and management work that currently takes place at schools prevents them from having time to prepare themselves, develop creativity and implement what they consider essential to do. It's just a wish list.

Internationally, the situation is very similar, as similar studies have shown: Pooley and O'Connor (2000), Khalid (2003), Christenson (2004), McKenzie (2005), Chrobak et al. (2006), Daskolia, Dimos & Kampylis (2012), Blanchet-Cohen & Reilly (2013), Almeida (2015), Avery & Norden (2015), Brunnquell, Brunstein & Jaime (2015) or Green, Medina-Jerez & Bryant (2015).

Therefore, this new EE proposal should be integrated into a process of transformation of both one's education and education systems, again in the context of a new society. It would contribute to the necessary restructuring of what we understand by education, a process that would be precisely fed back through innovative teaching and learning models in which the formation of future professionals of education is essential.

8.6 Conclusions

The only solution to all social and environmental problems we face today is a real sustainable, supportive, ecological development (in the true, scientific sense of the term), allowing for a complete transformation of our morals and our ethics, where nature and other living beings have their place, definitely banishing radical anthropocentrism of our world view and life. And this will be possible primarily through education: raising awareness thereon for a very near future. And one of the main actions addressing climate change lies, of course, on an EE transformed into an education in environmental values with a clear objective: the educational and informational nature. Given the present situation, it's more necessary than ever to act in such a complex and serious problem.

We advocate a new model of interdisciplinary and multidimensional environmental education with a global and integrated approach taking as its starting point the common values shared by all human beings. This will trigger criticism of what the phenomenon of globalization is and what, in essence, is our civilization like today. Environmental problems can not be separated from social and vice versa. They form a whole. Progress will only come through dialogue in all possible dimensions: social, ethical, moral, cultural, scientific, technological, economic, religious, etc. Based on those common elements (and all human beings have the same basic needs) it will be possible to build a global EE application in all educational systems in the world (Gómez Galán 2008). In this model, Climate Change Education (CCE) is an inseparable part of EE.

At present, multiple pedagogical and didactic proposals coexist, all based on the various existing environmental ethics (anthropocentric, zoocentric, biocentric and ecocentric). That is why today it is practically unfeasible to achieve common goals if we do not unleash a change. The key is to get the most positive and efficient of all of them and, in a dialogue process, conduct a construct that allows us reflecting on the major problems of the world in which we live, our main needs, which are the most appropriate strategies to improve it, and considering that we are one with nature (not different realities, what happens to it will happen to us), including, without further delay, the group of creatures like us, children of the Earth, within our ethical sphere. Thus, while fighting against poverty and social inequality we'll increase human welfare and we respect the environment, in capital letters, with all that it entails, as we have mentioned.

Focusing on a unified educational perspective is the very basis of the social development of mankind. We urgently need to change the current educational schema. Critical thinking must be based on knowledge. Teachers trained to do so, with a holistic and not just technical training, teachers who give preeminence to educational processes dominated by the most transcendent values will be essential for this purpose. The EE, as an essential part of what we must understand as education, may be one of the main engines that power this change, this transformation. No doubt, it is dealing with one of the most pressing challenges facing humanity today and even brings into play not only our development but even our survival.

9

APPENDIX: Using the SPSS Software to Determine Significant Statistical Difference between Two Groups of Data and Statistical Power

Mylord Reyes Tosta

Scientific Research Services, LLC, MI, United States

9.1 Introduction

The power of a statistical test is the probability that the test will reject the null hypothesis, when the null hypothesis is false. That is, the probability of not making a type II error or decision of false negatives (Cohen, 1988). That is, it represents the ability of a test to detect statistically significant differences or associations of a certain magnitude (Díaz & Fernández, 2003). When the power increases, the chances of a Type II error decreases. The probability of a Type II error refers to the false negative rate (β). Therefore, power is $1 - \beta$, where β (beta) is the Type II error also known as sensitivity. Also, a power analysis can be used to calculate the minimum sample size required, and to calculate a minimal effect given sample size. In addition, the concept of power is used to make comparisons between different methods of statistical analysis, for example, between a parametric test and a nonparametric test with the same hypothesis (Cohen, 1988).

Factors that influence statistical power of any study. The factors influencing the statistical power of a study will depend on each particular situation of a given study. The following four factors always influence the statistical power of a test.

1. The sample size used to detect the effect. It determines the amount of sampling error inherent to the test result. It is difficult to detect an effect on small samples. It is possible by increasing the sample size to obtain a higher power.

2. The magnitude of the interest effect in the population. It can be quantified in terms of the effect size. Where there is a higher power, there is a greater effect.
3. The level of statistical significance used in the test. A level of statistical significance is a statement of how unlikely may be a result if the null hypothesis is true to being considered significant. In other words, how much you're willing to take the risk of reaching a wrong conclusion. The most commonly used criteria are the odds of 0.05 (5%, 1 in 20), 0.01 (1%, 1 in 100) and 0001 (0.1%, 1 in 1000). If the criterion is 0.05, the probability of obtaining the observed effect when the null hypothesis is true must be less than 0.05 and so on. If a 5% significance level is used, known as alpha, or the probability of committing a Type I error, it means has a 95% confidence level. An easy way to increase the power of a test is to perform a less conservative test using a higher level of significance. This increases the probability of rejecting the null hypothesis. That is, obtaining a statistically significant result when the null hypothesis is false. Thus, the risk of committing a Type II error is reduced. But it also increases the risk of obtaining a statistically significant result rejecting the null hypothesis when the null hypothesis is true. In this case, the risk of making a Type I error is increased (Cohen, 1988).
4. The response variability or standard deviation of the study. Thus, if the response variability is greater, it will be harder to detect differences between the groups being compared and the study statistical power will be lower. It is recommended to study equivalent groups.

9.2 Results and Interpretation

The proper statistical power analysis of a study, which is ultimately the ability of the study to find differences, if indeed any, is a fundamental step both in the design phase and in the interpretation and discussion of results. At the time of design, therefore, the minimum magnitude of the difference or association that is considered relevant and the desired statistical power for the study should be established and the sample size should be calculated (Gordon, Finch, Nothnage & Ott, 2002). Whether the findings are statistically significant, or not, the estimation of confidence intervals can also facilitate interpretation of the results in terms of magnitude and significance, giving an idea of the precision with which the estimate has been made, of the magnitude, and of the effect direction. Thus, confidence intervals allow us to have an idea about the statistical power of a study, and therefore, of the credibility of the absence of significant findings (Díaz & Fernández, 2003). Consider the following points to interpret the results of statistical power:

a. According to Myoung (2003), the appropriate standard of power by most researchers is 0.80
b. Type II error = 0.20
c. A statistical power = 0.80 indicates the likelihood of saying that there is a relationship, difference, or gain. These are the odds that confirm our theory correctly.
d. A statistical power of 0.80 indicates that 80 out of 100 times when there is an effect, we are going to say that there is (Myoung, 2003).
e. If the statistical power is greater than 0.80, the power is more dominant.

Table 9.1 shows an illustration of the correct or incorrect conclusions that can be reached, depending on the rejection or acceptance of the null hypothesis. Where α = probability of making a Type I error and β = probability of making a Type II error.

In order to analyze the behavior of statistical power when alpha changes in a study, here there are two examples. In the first example, it has the results of a pre-test and post-test performed to a group of 48 subjects. They are the grades obtained by students on a scale from 1 to 100. The Table 9.2 shows a mean for the pre-test of 37.19 with a standard deviation of 17.43. While

Table 9.1 Statistical test of hypothesis contrast

		Possible Conclusions	
		Test Result	
		Association or significant difference. The null hypothesis is rejected	Association or no significant difference. The null hypothesis is accepted
Scenarios	There is a relationship or difference H0 false	No error $(1-\beta)$ Statistical Power	Type II error β Beta
	There is not a relationship or difference H0 true	Type I error α Significance level	No error $(1-\alpha)$ Confidence level

Paired Samples Statistics

Table 9.2 Mean and standard deviation of the pre-test and post-test

		Mean	N	Std. Deviation	Std. Error Mean
Pair 1	Pretest	37.19	48	17.432	2.516
	Posttest	77.88	48	12.261	1.770

the post-test, has a mean of 77.88 with a standard deviation of 12.26 for a difference between the means of 40.69 in this first example. Clearly, it can be noted that the results of the post-test outperform to the results of the pre-test. In the paired hypothesis test an alpha of 5% was used and obtained a p = .000 as shown in Table 9.3. With the obtained value (p = 0.000) it can be concluded that there exists a statistically significant difference between the results of the pre-test and post-test. So, the null hypothesis was rejected. With this result the statistical power of the test shown in Table 9.4 was calculated. A statistical power of 1.000 was obtained, which means a maximum statistical power. Therefore, it can be said with 100% certainty that there is really a statistically significant difference between the means. For all statistical calculations the SPSS version 20 Software was used.

Paired Samples Test

Table 9.3   Paired hypothesis testing with an alpha of 5%

		Paired Differences					t	Df	Sig. (2-tailed)
		Mean	Std. Deviation	Std. Error Mean	95% Confidence Interval of the Difference				
					Lower	Upper			
Pair 1	Pretest Posttest	−40.688	20.133	2.906	−46.533	−34.842	−14.002	47	.000

Tests of Within-Subjects Contrasts

Table 9.4   Statistical power of the paired hypothesis testing with an alpha of 5%

Measure: MEASURE_1

Source	Factor1	Type III Sum of Squares	Df	Mean Square	F	Sig.	Noncent. Parameter	Observed Power[a]
factor1	Linear	39731.344	1	39731.344	196.046	.000	196.046	1.000
Error (factor1)	Linear	9525.156	47	202.663				

a. Computed using alpha = .05

The same data from the first example was used to calculate the statistical power using an alpha of 1%. As shown in Tables 9.5 and 9.6, both the value of p in the paired hypothesis testing as the value of statistical power are exactly the same as when an alpha of 5% was used.

In the second example, it has the results of a pre-test and post-test performed to a group of 30 subjects. They are the grades obtained by the

Paired Samples Test

Table 9.5   Paired hypothesis testing with an alpha of 1%

							t	Df	Sig. (2-tailed)
			Mean	Std. Deviation	Std. Error Mean	99% Confidence Interval of the Difference			
						Lower	Upper		
Pair 1	Pretest – Posttest		−40.688	20.133	2.906	−48.489	−32.886	−14.002	47	.000

Tests of Within-Subjects Contrasts

Table 9.6   Statistical power of the paired hypothesis testing with an alpha of 1%

Measure: MEASURE_1

Source		factor1	Type III Sum of Squares	Df	Mean Square	F	Sig.	Noncent. Parameter	Observed Power[a]
factor1		Linear	39731.344	1	39731.344	196.046	.000	196.046	1.000
Error (factor1)		Linear	9525.156	47	202.663				

a. Computed using alpha = .01

students on a scale from 1 to 100. Table 9.7 shows a mean for the pre-test of 62.20 with a standard deviation of 19.04. While the post-test, has a mean of 64.53 with a standard deviation of 16.88 for a difference between the means of 2.33 in this second example. Clearly. it can be noted that the results of the post-test outperform the results of the pre-test. In the paired hypothesis testing was used an alpha of 5% and was obtained a p = .012 as shown in Table 9.8. With the value obtained (p = 0.012) it can be concluded that there exists a statistically significant difference between the results of the pre-test and post-test. Therefore, the null hypothesis was rejected. With this result was calculated the statistical power of the test shown in Table 9.9. A statistical power of .738 was obtained, which means a weak statistical power. Therefore, it can be said

Paired Samples Statistics

Table 9.7   Mean and standard deviation of the pre-test and post-test

		Mean	N	Std. Deviation	Std. Error Mean
Pair 1	Pretest	62.20	30	19.038	3.476
	Posttest	64.53	30	16.882	3.082

Paired Samples Test

Table 9.8   Paired hypothesis testing with an alpha of 5%

		Paired Differences					t	df	Sig. (2-tailed)
	Mean	Std. Deviation	Std. Error Mean	95% Confidence Interval of the Difference					
				Lower	Upper				
Pretest Posttest	−2.333	4.759	.869	−4.110	−.556		−2.686	29	.012

Tests of Within-Subjects Contrasts

Table 9.9   Statistical power of the paired hypothesis testing with an alpha of 5%

Measure: MEASURE_1

Source	Factor1	Type III Sum of Squares	Df	Mean Square	F	Sig.	Noncent. Parameter	Observed Power[a]
factor1	Linear	81.667	1	81.667	7.213	.012	7.213	.738
Error (factor1)	Linear	328.333	29	11.322				

a. Computed using alpha = .05

with only 74% certainty that there is really a statistically significant difference between the means.

The same data of the second example was used to calculate the statistical power using an alpha of 1%. As shown in Table 9.10, the result in the paired hypothesis testing with an alpha of 1% was p = .012. This is equal to that obtained with an alpha of 5%. The statistical power, while using an alpha of 1%, that is shown in Table 9.11 was 0.483. This value is totally different to that

Paired Samples Test

Table 9.10   Paired hypothesis testing with an alpha of 1%

			Paired Differences					t	Df	Sig. (2-tailed)
		Mean	Std. Deviation	Std. Error Mean	99% Confidence Interval of the Difference					
					Lower	Upper				
Pair 1	Pretest − Posttest	−2.333	4.759	.869	−4.728	.061		−2.686	29	.012

Tests of Within-Subjects Contrasts

Table 9.11 Statistical power of the paired hypothesis testing with an alpha of 1%

Measure: MEASURE_1

Source	Factor1	Type III Sum of Squares	Df	Mean Square	F	Sig.	Noncent. Parameter	Observed Power[a]
factor1	Linear	81.667	1	81.667	7.213	.012	7.213	.483
Error (factor1)	Linear	328.333	29	11.322				

a. Computed using alpha = .01

obtained with an alpha of 5% (.738). This statistical power result indicates that only can be said with 48% certainty that there exists a statistically significant difference between the means of the pre-test and post-test.

In light of the findings obtained, in the first example there was not difference in the statistical power when an alpha of 5% and alpha of 1% was used. While in the second example, when an alpha of 5% was used and it changed by an alpha of 1%, the value of the statistical power of the test decreased dramatically. This phenomenon is mainly due to two factors. The effect size is greater in the first example because the difference between the means was higher (40.69) compared to the difference of the second example which was 2.33. The effect size observed for the first example was .964 as shown in Table 9.12, which represents a large effect size. Finally, the sample in the first example was higher (48) compared to the sample of the second example (30). It can be concluded that in the second example the study did not have sufficient statistical power to ensure that there are significant differences between the pre-test and post-test. This indicates that in this study, the sample size should be increased. Finally, one might also point out that reducing the alpha level or type I error, if there is not a significant effect size and an appropriate sample size, the statistical power will fall irremediably.

Tests of Between-Subjects Effects

Table 9.12 The effect size observed for the first example

Measure: MEASURE_1

Transformed Variable: Average

Source	Type III Sum of Squares	Df	Mean Square	F	Sig.	Partial Eta Squared
Intercept	317745.094	1	317745.094	1263.303	.000	.964
Error	11821.406	47	251.519			

References

[1] Aikman, S. & Unterhalter, E. (Eds.). (2005). *Beyond Access: Transforming Policy and Practice for Gender Equality in Education.* London: Oxfam.

[2] Almeida, S. C. (2015). Teacher Education and Environmental Education. In S. C. Almeida (ed.). *Environmental Education in a Climate of Reform* (pp. 101–106). Dordrecht: Sense Publishers.

[3] Alonso, F., Manrique, D. & Vines, J. M. (2009). A Moderate Constructivist E-Learning Instructional Model Evaluated on Computer Specialists. *Computers & Education*, 53 (1), 57–65.

[4] Alvarado, M. & Ferguson, B. (1983). The Curriculum, Media Studies and Discursivity. *Screen*, 24, 3, 20–21.

[5] Alvarado, M., Gutch, R. & Wollen, T. (1987). *Learning the Media: An Introduction to Media Teaching.* London: Macmillan.

[6] Ander-Egg, E. (2003). *Métodos y Técnicas de Investigación Social IV: Técnicas para la recogida de Datos e Información.* Buenos Aires – México: Grupo Editorial Lumen Humanitas.

[7] Anderson, G. & Arsenault, N. (2001). *Fundamentals of Educational Research.* London: Routledge.

[8] Apel, H. (2007). Education for Sustainable Development. *Leibniz Perspectives*, 2, 49–51.

[9] Ardoin, N. M., Clark, C. & Kelsey, E. (2013). An Exploration of Future Trends in Environmental Education Research. *Environmental Education Research*, 19 (4), 499–520.

[10] Avery, H. & Norden, B. (2015). Within, above, between or outside?: ESD in Teacher Training: Implications of Various Institutional Constructions. In *World Environmental Education Congress, 8th WEEC*, Gothenburg: WEEC.

[11] Baek, J. (2012). *The Climate Change Education Evidence Base.* Washington D. C.: American Geophysical Union.

[12] Balzani, V. & Armaroli N. (2010). *Energy for a Sustainable World: From the Oil Age to a Sun-Powered Future.* Weinheim: Wiley-VCH Verlag.

[13] Banks, J. A. (2015). *Cultural Diversity and Education*. London: Routledge.

[14] Barrett, E. & Lally, V. (1999). Gender Differences in an On-Line Learning Environment. *Journal of Computer Assisted Learning*, 15 (1), 48–60.

[15] Barth, M., Godemann, J., Rieckmann, M. & Stoltenberg, U. (2007). Developing Key Competencies for Sustainable Development in Higher Education. *International Journal of Sustainability in Higher Education*, 8 (4), 416–430.

[16] Bazeley, P. (2013). *Qualitative Data Analysis: Practical Strategies*. Thousand Oaks, CA: Sage.

[17] Beatty, A. S., Feder, M. & Storksdieck, M. (Eds.). (2014). *Climate Change Education*. Washington D. C.: National Academies Press.

[18] Bekoff, M. & Goodall, J. (2003). *The Ten Trusts*. New York: Harper Collins.

[19] Bellard, C., Bertelsmeier, C., Leadley, P., Thuiller, W. & Courchamp, F. (2012). Impacts of Climate Change on the Future of Biodiversity. *Ecology Letters*, 15 (4), 365–377.

[20] Bereiter, C. & Scardamalia, M. (1989). Intentional Learning as a Goal of Instruction. In L. B. Resnick (ed.), *Knowing, Learning, and Instruction: Essays in Honor of Robert Glaser* (pp. 361–392). Hillsdale: Lawrence Erlbaum Associates.

[21] Bergman, M. (Ed.). (2008). *Advances in Mixed Methods Research*. Los Angeles: Sage Publications.

[22] Berkman, M. & Plutzer, E. (2010). *Evolution, Creationism, and the Battle to Control America's Classrooms*. Cambridge: Cambridge University Press, 2010.

[23] Bernard, H. R. & Bernard, H. R. (2012). *Social Research Methods: Qualitative and Quantitative Approaches*. Thousand Oaks, CA: Sage.

[24] Berríos, R. & Lucca, N. (2003). *Investigación Cualitativa: Fundamentos, Diseños y Estrategias*. Cataño, PR: SM.

[25] Berríos, R. & Lucca, N. (2013). *Investigación Cualitativa: Una Perspectiva Transdisciplinaria*. Cataño, PR: SM.

[26] Berryman, T. & Sauvé, L. (2016). Ruling Relationships in Sustainable Development and Education for Sustainable Development. *The Journal of Environmental Education*, 47 (2), 104–117.

[27] Bianchini, D. M. (2008). Contribuições para o Trabalho com Valores em Educação Ambiental. *Ciência & Educação*, 14 (2), 295–306.

[28] Black, M. C., Basile, K. C., Breiding, M. J., Smith, S. G., Walters, M. L., Merrick, M. T., Chen, J. & Stevens, M. R. (2011). *The National Intimate Partner and Sexual Violence Survey (NISVS): 2010 Summary Report.* Atlanta, GA: National Center for Injury Prevention and Control, Centers for Disease Control and Prevention.

[29] Blanchet-Cohen, N. & Reilly, R. C. (2013). Teachers' Pers-pectives on Environmental Education in Multicultural Contexts: Towards Culturally-Responsive Environmental education. *Teaching and Teacher Education*, 36, 12–22.

[30] Bless, C. & Higson-Smith, C. (2004). *Fundamentals of Social Research Methods*. Lusaka: Juta.

[31] Blum, N., Nazir, J., Breiting, S., Goh, K. C. & Pedretti, E. (2013). Balancing the Tensions and Meeting the Conceptual Challenges of Education for Sustainable Development and Climate Change. *Environmental Education Research*, 19 (2), 206–217.

[32] Boakye, C. (2015). Climate Change Education. *SAGE Open*, 5 (4), 1–10.

[33] Bogdan, R. & Biklen, S. (1992). Qualitative *Research for Education: An Introduction to Theory and Methods* (2nd Ed.). Boston, London, Toronto, Sydney, Tokyo, Singapore: Allyn and Bacon.

[34] Bonnett, M. (1999). Education for Sustainable Development: A Coher-ent Philosophy for Environmental Education? *Cambridge Journal of Education*, 29 (3), 313–324.

[35] Bonnett, M. (2002). Education for Sustainability as a Frame of Mind, *Environmental Education Research*, 8 (1), 9–20.

[36] Bordens, K. S. & Abbott, B. B. (2011). *Research Design and Methods: A Process Approach* (8th ed.). New York, NY: McGraw-Hill.

[37] Bostad, I. & Fisher, A. D. (2016). Curriculum and Social Change in Education for a Sustainable Future?. In Z. Babaci-Wilhite (ed.) *Human Rights in Language and STEM Education* (pp. 71–90). Dordrecht: Sense Publishers.

[38] Bowman, H. & Burden, T. (2002). Ageing, Community Adult Educa-tion, and Training. *Education and Ageing*, 17 (2–3), 147–167.

[39] Branisa, B., Klasen, S. & Ziegler, M. (2013). Gender Inequality in Social Institutions and Gendered Development Outcomes. *World Development*, 45, 252–268.

[40] Brewer, J. & Hunter, A. (2006). *Foundations of Multimethod Research: Synthesizing Styles*. Thounsand Oaks: Sage Publications.

[41] Brigandt, I. (2013). Intelligent Design and the Nature of Science: Philosophical and Pedagogical Points. In K. Kampourakis (ed.) *The Philosophy of Biology* (pp. 205–238). Dordrecht: Springer.

[42] Brown (2006). Still Subversive after all These Years: The Relevance of Feminist Therapy in the Age of Evidence-Based Practice. *Psychology of Women Quarterly*, 30 (1), 15–24.

[43] Brumfiel, G. (2005). Intelligent Design: Who Has Designs on Your Students' Minds? *Nature*, 434, 1062–1065.

[44] Brunnquell, C., Brunstein, J. & Jaime, P. (2015). Education for Sustainability, Critical Reflection and Transformative Learning: Professors' Experiences in Brazilian Administration Courses. *International Journal of Innovation and Sustainable Development*, 9 (3–4), 321–342.

[45] Bryman, A. (2004). *Social Research Methods*. Oxford: Oxford University Press.

[46] Buckingham, D. & Sefton-Green J. (1994). *Cultural Studies Goes to School. Reading and Teaching Popular Media*. London: Taylor and Francis.

[47] Buckingham, D. & Sefton-Green, J. (2003) Gotta Catch em All: Structure, Agency and Pedagogy in Children's Media Culture. *Media, Culture and Society*, 25 (3), 379–399.

[48] Buckingham, D. (2003). *Media Education. Literacy, Learning and Contemporary Culture*. Oxford: Polity Press.

[49] Buckingham, D. (2005). *The Media Literacy of Children and Young People: A Review of the Literature*. London: Centre for the Study of Children Youth and Media Institute of Education – University of London.

[50] Burns, R. B. (2000). *Introduction to Research Methods*. London: Sage.

[51] Caduto, M. (1983). A Review of Environmental Values Education. *Journal of Environmental Education*, 14, 13–21.

[52] Caduto, M. (1985). A Teacher Training Model and Educational Guidelines for Environmental Values Education, *Journal of Environmental Education*, 16, 30–34.

[53] Calder, W. & Clugston, R. (2005). Editorial: Education for a Sustainable Future. *Journal of Geography in Higher Education*, 29 (1), 7–12.

[54] Campos, A. (2009). *Métodos Mixtos de Investigación: Integración de la Investigación Cuantitativa y la Investigación Cualitativa*. Investigar el Magisterio. Colombia.

[55] Caruth, G. D. (2013). Demystifying Mixed methods research Design: A Review of the Literature. *Melvana International Journal of Education*, 3 (2), 112–122.

[56] Castillo Espitia, E. (2000). La Fenomenología Interpretativa como Alternativa Apropiada para Estudiar los Fenómenos Humanos. Revista Investigación y Educación en Enfermería. 27 (1), pp. 27–35.

[57] Castro, I. (2004). *La Pareja Actual Transición y Cambios*. Buenos Aires: Lugar Editorial.

[58] Cavalcanti, T. & Tavares, J. (2016). The Output Cost of Gender Discrimination: A Model – Based Macroeconomics Estimate. *The Economic Journal*, 126, 109–134.

[59] CEPAL (2013). *Observatorio para la Igualdad de Género de América Latina y el Caribe, Informe Anual 2012*. Santiago de Chile: Comisión Económica para América Latina y el Caribe.

[60] Charmaz, K. (2006). *Constructing Grounded Theory. A Practical Guide through Qualitative Analysis*. London: Sage.

[61] Chávez Tortolero, M. (2005). La Ética Ambiental como Reflexión en el Marco de la Educación en Ciencias y en Tecnología: Hacia el Desarrollo de la Conciencia de la Responsabilidad. *Educere*, 27, 1–8.

[62] Chisamya, G., DeJaeghere, J., Kendall, N. & Khan, M. A. (2012). Gender and Education for All: Progress and Problems in Achieving Gender Equity. *International Journal of Educational Development*, 32 (6), 743–755.

[63] Choi, H. J. & Johnson, S. D. (2005). The Effect of Context-Based Video Instruction on Learning and Motivation in Online Courses. *The American Journal of Distance Education*, 19 (4), 215–227.

[64] Christenson, M. A. (2004). Teaching Multiple Perspectives on Environmental Issues in Elementary Classrooms: A Story of Teacher Inquiry. *Journal of Environmental Education*, 35 (4), 3–17.

[65] Chrobak, R. et al. (2006). Una Aproximación a las Motivaciones y Actitudes del Profesorado de Enseñanza Media de la Provincia de Neuquén sobre Temas de Educación Ambiental. *Revista Electrónica de Enseñanza de las Ciencias*, 5 (1), 31–50.

[66] Cisneros-Cohernour, E. (2012). Case Studies and Validity. (Chapter 33. In Charles Secolsky and Brian Denison (2012 Ed.). *Handbook on Measurement, Assessment, and Evaluation in Higher Education*. New York and London: Routledge.

[67] Clark, R. & Sugrue, B. (1990). North American Disputes about Research on Learning from Media. *International Journal of Educational Research*, 14, 6, 507–520.

[68] Clark, R. (1985). Confounding in Educational Computing Research. *Journal of Educational Computing Research*. 1 (2), 137–148.

[69] Clark, R. E. & Salomon, C. (1977). Reexamining the Methodology of Research on Media and Technology in Education. *Review of Educational Research*, 47 (1), 99–120.

[70] Clark, R. E. & Surgrue, B. M. (1988). Research on Instructional Media, 1978–1988. In D. P. Eli, B. Broadbent & R. K. Wood (eds.). *Educational Media and Technology Yearbook*. (vol. 14, pp. 19–36). Englewood: Libraries Unlimited.

[71] Clasificaciones Carnegie (2009). Clasificaciones de la Comisión Carnegie para las Instituciones de Educación Superior. Recuperado de http://74.125.67.132/translate_c?hl=es&sl=en&u=http://www.carnegie foundation.org/class

[72] Clément, P. & Caravita S. (2014). Education for Sustainable Development: International Surveys on Conceptions and Postures of Teachers. C. Bruguière, Tiberghien, & P. Clément. *Topics and Trends in Current Science Education* (pp. 175–192) Dordrecht: Springer.

[73] Cocks, S. & Simpson, S. (2015). Anthropocentric and Ecocentric: An Application of Environmental Philosophy to Outdoor Recreation and Environmental Education. *Journal of Experiential Education*, Preprint. February 19, 2015.

[74] Cohen, J. (1988). *Statistical Power Analysis for the Behavioral Sciences*. (2nd ed.). New Jersey: Lawrence Erlbaum.

[75] Cohen, L. & Manion, L. (1990). *Métodos de Investigación Educativa*. Madrid: Editorial La Muralla.

[76] Cohen, L., Manion, L. & Morrison, K. (2000). *Research Methods in Education*. London: Routledge.

[77] Cohen, L., Manion, L. & Morrison, K. (2013). *Research Methods in Education*. London: Routledge.

[78] Complexo Hospitalario Juan Canalejo. Significancia Estadística. *Cad Aten Primaria*, 10, 59–63.

[79] Connell, R. W. (2014). *Gender and Power: Society, the Person and Sexual Politics*. New York: John Wiley & Sons.

[80] Conrad, C. & Serlin, R. C. (Eds.) (2011). *The Sage Handbook for Research in Education: Pursuing Ideas as the Keystone of Exemplary Inquiry* (2nd ed.). Thousand Oaks, CA: Sage.

[81] Corbin, J. & Strauss, H. (2008). *Basic Qualitative Research: Techniques and Procedures for Developing Grounded Theory*. Sage Publications, Inc.

[82] Corey, G. (2013). *Theory and Practice of Counseling and Psychotherapy*. Monterey: Brooks Cole.

[83] Creswell, J. & Plano, C. V. (2011). *Designing and Conducting Mixed Methods Research* (2nd Ed.). Sage Publications: Thounsand Oaks.

[84] Creswell, J. (2003). *Research Design: Qualitative, Quantitative, and Mixed Methods Approaches*. Thousand Oaks, CA: Sage.

[85] Creswell, J. (2007). *Qualitative Inquiry & Research Design: Choosing Among Five Approaches* (2nd Ed.). Thousand Oaks: Sage Publications.

[86] Creswell, J. (2009). Editorial: Mapping the field of mixed methods research. *Journal of Mixed Methods Research*, 3 (2), 95–108.

[87] Creswell, J. W. (2012). *Educational Research: Planning, Conducting, and Evaluating Quantitative and Qualitative Research*. Boston: Pearson.

[88] Creswell, J. W. (2012). *Qualitative Inquiry and Research Design: Choosing Among Five Approaches*. Thousand Oaks, CA: Sage.

[89] Criticos, C. (2000). La Educación para los Medios: Un Compromiso con la Democracia. *Tabanque*, 14, 35–42.

[90] Cronbach, L. J. & Snow, R. E. (1977). *Aptitudes and Instructional Methods: A Handbook for Research on Interactions*. New York: Irvington Publishers.

[91] Crutzen, P. J. & Stoermer, E. F. (2000). The Anthropocene. *Global Change Newsletter*, 41, 17–18.

[92] Daskolia, M., Dimos, A. & Kampylis, P. G. (2012). Secondary Teachers' Conceptions of Creative Thinking within the Context of Environmental Education. *International Journal of Environmental and Science Education*, 7 (2), 269–290.

[93] DeJaeghere, J. & Wiger, N. P. (2013). Gender Discourses in an NGO Education Project: Openings for Transformation Toward Gender Equality in Bangladesh. *International Journal of Educational Development*, 33, 557–565.

[94] Dembski, W. A. (2002). *Intelligent Design: The Bridge between Science & Theology*. Downer's Grove: InterVarsity Press.

[95] Denzin, N. & Lincoln, Y. (Eds.). (2008). *Collecting and Interpreting Qualitative Materials*. Thousand Oaks, CA: Sage.

[96] Denzin, N. (2009). *Qualitative Inquiry Under Fire: Toward a New Paradigm Dialogue*. Walnut Creek, California: Left Cost Press Inc.

[97] Díaz Nosty, B. (2009). Cambio Climático, Consenso Científico y Construcción Mediática. Los Paradigmas de la Comunicación para la Sostenibilidad. *Revista Latina de Comunicación Social*, 64, 99–119.

[98] Díaz, P. & Fernández, P. (2003). Cálculo del poder estadístico de un estudio. Unidad de Epidemiología Clínica y Bioestadística. Complejo Hospitalario-Universitario Juan Canalejo. A Coruña, España. *Cad Aten Primaria*, 10, 59–63.

[99] Diggle, P. & Chetwynd, A. (2011). *Statistics and Scientific Method: An Introduction for Students and Researchers*. New York, NY: Oxford University Press.

[100] DJS Research, Ltd. (2010). Exploratory Research. Recuperado de Market Research Company UK. Company Registration Number: 5494158.

[101] Dolan, K. (2008). Comparing Modes of Instruction: The Relative Efficacy of On-Line and In-Person Teaching for Student Learning. *PS: Political Science and Politics*, 41 (2), 387–391.

[102] Dooley, D. (2001). *Social Research Methods*. Englewood Cliffs, NJ: Prentice Hall.

[103] ECPI (1998). *Estatuto para la Corte Penal Internacional*. Roma: Corte Penal Internacional.

[104] Eisner, E. & Peshking., A. (Ed.) (1990). *Qualitative Inquiry in Education: The Continuing Debate*. Teachers College, Columbia University. New York.

[105] Ellis, H. (2005). Descriptive Cataloging Proficiencies among Beginning Students: A Comparison among Traditional-Class and Virtual-Class LIS Students. *Journal of Library & Information Services in Distance Learning*, 2 (2), 13–43.

[106] Escobar, J. & Ovalle, C. (2015). The Role of Bioethics in the Resolution of Environmental Conflicts. *Revista Colombiana de Bioética*, 10 (1), 65–85.

[107] Fairclough, N. (2005). Critical Discourse Analysis. *Marges Linguistiques*, 9, 76–94.

[108] Fankhauser, S. (2013). *Valuing Climate Change: The Economics of the Greenhouse*. London: Routledge.

[109] Ferkany, M. & Whyte, K. P. (2012). The Importance of Participatory Virtues in the Future of Environmental Education. *Journal of Agricultural and Environmental Ethics*, 25 (3), 419–434.

[110] Fernandez, G., Thi, T. T. M. & Shaw, R. (2014). Climate Change Education: Recent Trends and Future Prospects. In R. Shaw & Y.

Oikawa (eds.). *Education for Sustainable Development and Disaster Risk Reduction: Methods, Approaches and Practices* (pp. 53–74). Kyoto: Springer Japan.

[111] Fernández, N. (2008). *Education of Adults in the Institutions of Higher Education in Puerto Rico. Collaborative Agenda for Action.* San Juan: PR.

[112] Fien, J. (2003). Education for a Sustainable Future: Achievements and Lessons from a Decade of Innovation, from Rio to Johannesburg. *International Review for Environmental Strategies*, 4 (1), 5–20.

[113] Flick, U. (2004). Design and Process in Qualitative Research. In U. Flick, E. von Kardoff & I. Steinke (eds) *A Companion to Qualitative Research* (pp. 146–152). London: Sage.

[114] FRA (2014). *Violence Against Women: An EU-Wide Survey.* Bruselas: Agency for Fundamental Rights.

[115] Freebody, P. (2003). Qualitative *Research in Education: Interaction and Practice.* London, Thousand Oaks, New Delhi: Sage Publications.

[116] Freeman, F. N. (1924). *Visual Education, a Comparative Study of Motion Pictures and Other Methods of Instruction.* Chicago: University of Chicago Press.

[117] Freire, P. (1984). *Sobre Educação.* Río de Janeiro: Paz e Terra.

[118] Freire, P. (2006). Pedagogía de la Autonomía: Saberes Necesarios para la Práctica Educativa. Madrid: Siglo XXI.

[119] Freire, P. (2014). *Educação como Prática da Liberdade.* Río de Janeiro: Paz e Terra.

[120] Fuentes, L. Y. (2006). Género, Equidad y Ciudadanía: Análisis de las Políticas Educativas. *Nómadas*, 24, 22–35.

[121] Gay, L. R., Mills, G. E. & Airasian, P. W. (2011). *Educational Research: Competencies for Analysis and Applications.* New York, NY: Pearson.

[122] Gay, R. L., Mills, G. E. & Airasian, P. W. (2011). *Educational Research: Competencies for Analysis and Application.* Upper Saddle River: Merrill.

[123] Giroux, H. (1992). *Border Crossings: Cultural Workers and the Politics of Education.* New York: Routledge.

[124] Glaser, B. & Strauss, A. (1967). *The Discovery of Grounded Theory: Strategies For Qualitative Research.* New York: Aldine de Gruyter.

[125] Gómez Galán, J. & Mateos, S. (2002a). Análisis Teórico de las Necesidades Educativas en Tecnologías y Medios de Comunicación del Alumnado de Educación Primaria y Secundaria. *Polyphonía*, 13, 44–57.

[126] Gómez Galán, J. & Mateos, S. (2002b). Retos Educativos en la Sociedad de la Información y la Comunicación. *Revista Latinoamericana de Tecnología Educativa*, 1 (1), 9–23.

[127] Gómez Galán, J. & Mateos, S. (2004). Design of Educational Web Pages. *European Journal of Teacher Education.* 27 (1), 99–104.

[128] Gómez Galán, J. (1999). *Tecnologías de la Información y la Comunicación en el Aula.* Madrid: Seamer.

[129] Gómez Galán, J. (2003). *Educar en Nuevas Tecnologías y Medios de Comunicación.* Badajoz-Seville: F.E.P.

[130] Gómez Galán, J. (2005). Animal Rights Education: An Initiative at University. In N. Jukes & S. Martinsen (eds.). *Alternatives in the Mainstream: Innovations in Life Science Education and Training.* Proceedings of the 2nd InterNICHE Conference. Leicester-Oslo: InterNICHE.

[131] Gómez Galán, J. (2007). Los Medios de Comunicación en la Convergencia Tecnológica: Perspectiva Educativa. *Comunicación y Pedagogía*, 221, 44–50.

[132] Gómez Galán, J. (2008). *Valores Medioambientales en la Educación: Situación del Futuro Profesorado de Extremadura ante la Ecología y el Cambio Climático.* [Premios Nacionales de Investigación e Innovación Educativa]. Madrid: Ministerio de Educación y Ciencia.

[133] Gómez Galán, J. (2010). *Educación, Protección Animal y Bioética.* Badajoz: UEX.

[134] Gómez Galán, J. (2011). New Perspectives on Integrating Social Networking and Internet Communications in the Curriculum. *eLearning Papers*, 26, 1–7.

[135] Gómez Galán, J. (2014a). El Fenómeno MOOC y la Universalidad de la Cultura: Las Nuevas Fronteras de la Educación Superior. *Revista de Curriculum y Formación del Profesorado*, 18 (1), 73–91.

[136] Gómez Galán, J. (2014b). Educación y Globalización: Desarrollos Pedagógicos Innovadores en el Contexto de la Cultura Postmoderna. In J. C. Martínez Coll, *Educación, Cultura y Desarrollo* (pp. 34–51). Málaga: Universidad de Málaga.

[137] Gómez Galán, J. (2015a). Education against Global Warming and Climate Change. In J. Gómez Galán, (ed.) *Climate Change: A Multidisciplinary View.* Cupey: UMET Press.

[138] Gómez Galán, J. (2015a). Media Education as Theoretical and Practical Paradigm for Digital Literacy: An Interdisciplinary Analysis. *European Journal of Science and Theology*, 11 (3), 31–44.

[139] Gómez Galán, J. & Mateos, S. (2002). Retos Educativos en la Sociedad de la Información y la Comunicación. *Revista Latinoamericana de Tecnología Educativa*, 1 (1), 9–23.

[140] Gómez Heras, J. M. (1998). *Ética del Medio Ambiente*, Madrid: Tecnos.

[141] González Gaudiano, E. (2007). La Educación Ambiental de Cara a la Problemática Ambiental Global. *Ciencia UANL*, 10 (4), 425–432.

[142] González Pozuelo, F. (ed.), Gómez Galán, J., Pérez Rubio, J. A., Blanco, R., Rumbao, J. & Navareño, P. (2009). *Sexismo y Violencia de Género en la Población Escolar de Extremadura*. Badajoz: Universidad de Extremadura.

[143] González Raimundí, Z. (2009). *Granada y Melilla: una Experiencia Educativa Más Allá de la Frontera*. Ensayo publicado en página electrónica de Universidad Metropolitana, Cupey, PR.

[144] Gorard, S. & Taylor, C. (2004). *Combining Methods in Educational and Social Research*. London: Open University Press.

[145] Gorard, S. (2001) *Quantitative Methods in Educational Research: The Role of Numbers Made Easy*. London: Continuum.

[146] Gordon, D., Finch, S. J., Nothnagel, M. & Ott, J. (2002). Power and Sample Size Calculations for Case-Control Genetic Association Tests when Errors are Present: Application to Single Nucleotide Polymorphisms. *Hum Hered* 54, 22–23.

[147] Gratton-Lavoie, C. & Stanley, D. (2009). Teaching and Learning Principles of Microeconomics Online: An Empirical Assessment. *Journal of Economic Education*, 40 (1), 3–25.

[148] Green, C., Medina-Jerez, W. & Bryant, C. (2015). *Cultivating Environmental Citizenship in Teacher Education. Teaching Education*, 27 (2), 1–19.

[149] Greene, J. (2007). *Mixed Methods in Social Inquiry*. California: Jossey-Bass & Wiley.

[150] Gruen, L. (2011). *Ethics and Animals: An Introduction*. Cambridge: Cambridge University Press.

[151] Grunberg, L. (2004). Access to Gender–Sensitive Higher Education in Eastern and Central Europe. *European Education*. 4, 54–69.

[152] Grusky, D. B. & Weisshaar, K. R. (Eds.). (2001). *Social Stratification: Class, Race, and Gender in Sociological Perspective*. Boulder: Westview Press.

[153] Guba, E. (1990). *The Paradigm Dialog*. International Educational and Professional Publisher. Newbury Park, London, New Delhi: Sage Publications.

[154] Hackbarth, S. (1996). *The Educational Technology Handbook. Process and products for learning*. Englewood Cliffs: Educational Technology Publications.

[155] Halfacree, A. & Ellison, J. (2001). Education for Sustainable Development. *Education in Science*, 193, 14–15.

[156] Hall, S. (1977). Culture, the Media and the Ideological Effect. In J. Curran, M. Gurevitch & J. Woollacott (eds.) *Mass Communication and Society*. (pp. 315–348). London: Edward Arnold.

[157] Hammersley, M. (2008). *Questioning Qualitative Inquiry: Critical Essays*. Los Angeles: Sage Publications.

[158] Hargrave, C. P., Simonson, M. R. & Thompson, A. D. (1996). *Educational Technology: A Review of the Research*. Ames: Iowa State University.

[159] Hart, A. (1991). *Understanding the Media*. London & New York: Routledge.

[160] Hartas, D. (Ed.). (2015). *Educational Research and Inquiry: Qualitative and Quantitative Approaches*. New York, NY: Bloomsbury Publishing.

[161] Hartley, J. (1974). Programmed Instruction 1954–1974: A Review. *Programmed Learning and Educational Technology*, 11 (6), 278–291.

[162] Hartsell, B. (2006). Teaching toward Compassion: Environmental Values Education for Secondary Students. *Prufrock Journal*, 17 (4), 265–271.

[163] Hernández Sampieri R., Fernández Collado C. & Baptista, P. (2010). *Metodología de la Investigación*. México: Mc Graw Hill.

[164] Hernández Sampieri, R., Fernández Collado, C. & Baptista Lucío, P. (2006). *Metodología de la Investigación*. Mexico, D.F.: McGraw-Hill.

[165] Hernandez, C. & Mayur, R. (eds.). (2000). *Pedagogy of the Earth: Education for a Sustainable Future*. Mumbai: International Institute for Sustainable Future.

[166] Hickey, W. D. (2013). Intelligent Design in the Public School Science Classroom. *Interchange*, 44 (1–2), 25–29.

[167] Hourdequin, M. (2015). *Environmental Ethics: From Theory to Practice*. London: Bloomsbury Publishing.

[168] Houser, R. (2008). *Counseling and Educational Research: Evaluation and Application*. Thousand Oaks, CA: Sage.

[169] Hoy, W. K. & Adams, C. M. (2016). *Quantitative Research in Education* (2nd ed.). Thousand Oaks, CA: Sage.

[170] Human Rights Watch (2001). *Scared at School: Sexual Violence against Girls in South African Schools*. New York: Human Rights Watch.

[171] Hung, C. C. (2014). *Climate Change Education: Knowing, Doing and Being*. London: Routledge.

[172] Hurst, C. E. (2015). *Social Inequality: Forms, Causes, and Consequences*. London: Routledge.

[173] Imran, S., Alam, K. & Beaumont, N. (2014). Reinterpreting the Definition of Sustainable Development for a More Ecocentric Reorientation. *Sustainable Development*, 22 (2), 134–144.

[174] Inglehart, R. & Norris, P. (2003). *Rising Tide: Gender Equality and Cultural Change around the World*. Cambridge: Cambridge University Press.

[175] Jämsä, T. (2006). The Concept of Sustainable Education. In A. Pipere (Ed.), *Education & Sustainable Development: First Steps toward Changes* (Vol. 1, pp. 5–30). Daugavpils: Daugavpils University.

[176] Jandaghi, G. & Matin, H. Z. (2009). Achievement and Satisfaction in a Computer-assisted Versus a Traditional Lecturing of an Introductory Statistics Course. *Australian Journal of Basic and Applied Sciences*, 3 (3), 1875–1878.

[177] Jenkins, H., Clinton, K., Purushotma, R., Robison, A. J. & Weigel, M. (2006). *Confronting the Challenges of Participatory Culture. Media Education for the 21st Century*. London-Cambridge, MA: The MIT Press.

[178] Jiménez Fernández, C. (2011). Educación, Género e Igualdad de Oportunidades. *Tendencias Pedagógicas*, 18, 51–85.

[179] Johnson R. B. & Christensen, L. B. (2007). *Educational Research: Quantitative, Qualitative, and Mixed Approaches*. Thousand Oaks, CA: Sage.

[180] Johnson, B. & Christensen, L. (2000). *Educational Research: Quantitative and Qualitative Approaches*. New York, NY: Allyn & Bacon.

[181] Kalogirou, S. A. (2013). *Solar Energy Engineering: Processes and Systems*. Oxford: Academic Press.

[182] Kauchak, D., Krall, R. & Hemsath, K. (1978). Environmental Values Education: The Need for Education, not Indoctrination. *Journal of Environmental Education*, 10 (1), 19–22.

[183] Kaufman, G. (2000). Do Gender Role Attitudes Matters? Family Formation and Dissolution among Traditional and Egalitarian Men and Women. *Journal of Family Issues*, 21, 128–144.

[184] Khalid, T. (2003). Pre-service High School Teachers' Perceptions of Three Environmental Phenomena. *Environmental Education Research*, 9 (1), 35–50.

[185] Kieft, M., Rijlaarsdam, G. & Van den Bergh, H. (2008). An Aptitude Treatment Interaction Approach to Writing-to-Learn. *Learning and Instruction*, 18, 379–390.

[186] Kim, S. & Feldt, L. S. (2010). The Estimation of the IRT Reliability Coefficient and Its Lower and Upper Bounds with Comparisons to CTT Reliability Statistics. *Asia Pacific Education Review*, 11 (2), 179–188.

[187] Kincheloe, J. L. (2003) *Teachers as Researchers: Qualitative Inquiry as a Path to Empowerment*. London: Routledge.

[188] Kirchhoff, M. M. (2010). Education for a Sustainable Future. *Journal of Chemical Education*, 87 (2), 121–121.

[189] Klein, S. S., Richardson, B., Grayson, D. A., Fox, L. H., Kramarae, C., Pollard, D. S. & Dwyer, C. A. (Eds.). (2014). Handbook for Achieving Gender equity Through Education. London: Routledge.

[190] Knapp, C. E. (1983). A Curriculum Model for Environmental Values Education. *Journal of Environmental Education*, 14, 22–26.

[191] Knowles, M. S.; Holton, E. F.; Swanson, R. A. (2001). *Andragogía: El Aprendizaje de los Adultos*. México, D.F.: Oxford University Press.

[192] Knowlton, J. Q. (1964). A Conceptual Scheme for Audiovisual Field. *Bulletin of the School of Education, Indiana University*. 40 (3), 1–44.

[193] Knudson-Martin C. & Rankin Mahoney A. (1998). Language and Processes in the Construction of Equality in New Marriages. *Family Relations*, 47, 81–91.

[194] Kopnina, H. (2013). Evaluating Education for Sustainable Development (ESD): Using Ecocentric and Anthropocentric Attitudes toward the Sustainable Development (EAATSD) Scale. *Environment, Development and Sustainability*, 15 (3), 607–623.

[195] Kopnina, H. (2014). Revisiting Education for Sustainable Development (ESD): Examining Anthropocentric Bias through the Transition of Environmental Education to ESD. *Sustainable Development*, 22 (2), 73–83.

[196] Koya, C. F. (2013). *The Significance of Climate Change Education*. Fiji: EU-USP Climate Change Alliance Project.

[197] Kronlid, D. O. & Öhman, J. (2013). An Environmental Ethical Conceptual Framework for Research on Sustainability and Environmental Education. *Environmental Education Research*, 19 (1), 21–44.

[198] Lampard, R. & Pole, C. (2015). *Practical Social Investigation: Qualitative and Quantitative Methods in Social Research*. London: Routledge.

[199] Lassoe, J., Schnack, K., Breiting, S., Rolls, S., Feinstein, N. & Goh, K. C. (2009). *Climate Change and Sustainable Development: The Response from Education.* Copenhagen: International Alliance of Leading Education Institutes.

[200] Lehrer, R. & Randle, L. (1987). Problem Solving, Metacognition and Composition: The Effect of Interactive Software for First-Grade Children. *Journal of Educational Computing Research.* 3 (4), 409–427.

[201] Lichtman, M. (2006). Qualitative *Research in Education: A User's Guide.* Thounsand Oaks: Sage Publications.

[202] Lichtman, M. (Ed.) (2011). *Understanding and Evaluating Qualitative Educational Research.*

[203] Lincoln, Y. & Guba, E. (1985). *Naturalistic Inquiry.* Newbury Park, London, New Delhi: Sage Publications. Los Angeles: Sage Publications.

[204] Longworth, N. (2003). *El Aprendizaje a lo Largo de la Vida en la Práctica. Transformar la Educación en el siglo XXI.* México, D.F.: Paidós.

[205] Macpherson, C. C. (2013). Climate Change is a Bioethics Problem. *Bioethics,* 27 (6), 305–308.

[206] Markauskaite, L. (2006). Gender Issues in Preservice Teachers' Training: ICT Literacy and Online Learning. *Australasian Journal of Educational Technology,* 22 (1), 1–20.

[207] Martínez Ramos, L. M. & Tamargo López, M. (2003). *Género, Sociedad y Cultura.* Colombia: Publicaciones Gaviota.

[208] Maslow, A. H. (1991). *Motivación y Personalidad.* Madrid: Díaz de Santos, S. A.

[209] Masterman, L. (1990). *Teaching the Media.* New York: Routledge.

[210] Masterman, L. (1991). An Overview of Media Education in Europe. *Media Development.* 27 (1), 3–9.

[211] Masterman, L. (1995). *The Media Education Revolution.* Southhampton: Southhampton Media Education Group.

[212] Masterman, L. (1998). Foreword. In A. Hart (Ed.). *Teaching the Media: International Perspectives.* (pp. 7–9). Mahway, NJ: Lawrence Erlbaum Associates.

[213] Masterman, L. (2001). A Rationale for Media Education. In R. Kubey, (Ed.). *Media Literacy in the Information Age: Current Perspectives. Information and Behaviour.* New Brunswick: Transaction Publishers.

[214] Mateos, M. M. (1991). Teoría de la Comunicación. In *Tecnología de la Educación.* (p. 507). Madrid: Santillana.

[215] Matthews, B. & Ross, L. (2010). *Research Methods: A Practical Guide for the Social Sciences*. New York, NY: Pearson.

[216] Maykut, P. & Morehouse, R. (2003). *Beginning Qualitative Research: A Philosophic and Practical Guide*. London: Falmer.

[217] McKenney, S. E. & Reeves, T. C. (2012). *Conducting Educational Design Research*. New York, NY: Routledge.

[218] McKenzie, M. (2005). The "Post-Post Period" and Environmental Education Research. *Environmental Education Research*, 11 (4), 401–412.

[219] McKeown, R., Hopkins, C. A., Rizi, R. & Chrystalbridge, M. (2002). *Education for Sustainable Development Toolkit*. Knoxville: University of Tennessee.

[220] McMillan, J. H. & Schumacher, S. (2014). *Research in Education: Evidence-based inquiry*. Pearson Higher Ed.

[221] McNiff, J. & Whitehead, J. (2002) *Action Research: Principles and Practice*. London: Routledge.

[222] Merriam, S. (2009). *Qualitative Research: A Guide to Design and Implementation*. Jossey-Bass. A Wiley Company Imprint.

[223] Merriam, S. and Associates (2002). *Qualitative Research in Practice: Examples for Discussion and Analysis*. San Francisco, CA: Jossey-Bass. A Wiley Company Imprint.

[224] Mertens, D. (2005). *Research and Evaluation in Education and Psychology: Integrating Diversity with Quantitative, Qualitative, and Mixed Methods* (2nd Ed.). Thousand Oaks: Sage Publications.

[225] Miles, M. & Huberman, M. (1994). *Qualitative Data Analysis: An Expanded Sourcebook* (2nd Ed.). London, Thousand Oaks, New Delhi: Sage Publications.

[226] Minteer, B. (2012). *Refounding Environmental Ethics*. Philadelphia: Temple University Press.

[227] Mochizuki, Y. & Bryan, A. (2015). Climate Change Education in the Context of Education for Sustainable Development: Rationale and Principles. *Journal of Education for Sustainable Development*, 9 (1), 4–26.

[228] Mogensen, F. & Schnack, K. (2010). The Action Competence Approach and the "New" Discourses of Education for Sustainable Development, Competence and Quality Criteria. *Environmental Education Research*, 16 (1), 59–74.

[229] Moletsane, R. (2005). Gender Equality in Education in the Context of the Millenium Development Goals: Challenges and Opportunities for Women. *Convergence*, 28, 59–68.

[230] Montero, J. (2006). Feminismo: Un Movimiento Critico. *Intervención Psicosocial*, 15, 167–180.

[231] Morse, J. & Niehaus, L. (2009). *Mixed Method Design: Principles and Procedures*. California: Left Cost Press Inc.

[232] Mosconi, N. (2014). Escola Mista e Igualdade entre os Sexos no Contexto Frances. *Educaçao & Realidade*, 39, 221–239.

[233] Mosterín, J. (1995). *Los Derechos de los Animales*. Madrid: Dominós.

[234] Myoung, H. (2003). *Understanding the Statistical Power of a Test*. UITS Center for Statistical and Mathematical Computing. Indiana University.

[235] National Research Council (2002). *Scientific Research in Education*. Washington, DC: National Academy Press.

[236] Negroponte, N. (1995). *Being Digital*. New York: Alfred A. Knopf.

[237] Nouri, H. & Shahid, A. (2005). The Effect of Power Point Presentations on Student Learning and Attitudes, *Global Perspectives on Accounting Education*, 2 (1), 53–73.

[238] Núñez, M. (2009). Hacia Dónde Vamos en la Educación Virtual. *Presentation Transcript*. Instituto para el Desarrollo de la Enseñanza y el Aprendizaje en Línea. Recuperado de www.slideshare.net/.../hacia-donde-vamos-en-la-educacion-virtual

[239] ONU (1986). *Report of the Expert Group Meeting on Violence in the Family with Special Emphasis on its Effects on Women*. Vienna: UN/Division for the Advancement of Women.

[240] ONU (1992). *Report of the Commission on the Status of Women at its Thirty-Sixth Session (E/1992/24; E/CN.6/1992/13)*. New York: UN/Economic and Social Council.

[241] ONU (1993). *Expert Group Meeting on Measures to Eradicate Violence Against Women. Report (MAV/1993/1)*. Nuena York: UN/Division for the Advancement of Women, Department for Policy Coordination and Sustainable Development.

[242] ONU (2008): *Comisión de la Condición Jurídica y Social de la Mujer 52° Período de Sesiones. Conclusiones Convenidas sobre la Financiación a favor de la Igualdad entre los Géneros y el Empoderamiento de la Mujer*. New York: UN/ Economic and Social Council.

[243] Pachauri R. K. & Meyer, L. A. (eds.). *IPCC, 2014: Climate Change 2014. Synthesis Report. Contribution of Working Groups I, II and III to the Fifth Assessment Report of the Intergovernmental Panel on Climate Change*. Geneva: IPCC.

[244] Padilla, M. & Gómez-Galán, J. (2014). Análisis Discursivo de la Construcción y Deconstrucción de la Equidad: Un Estudio de Caso para su Aplicación en el Ámbito de la Educación para la Igualdad. *IJERI: International Journal of Educational Research and Innovation*, 1 (1), 14–28.

[245] PAHO (1993). *La Violencia contra las Mujeres y las Niñas: Análisis y Propuestas desde la Perspectiva de la Salud Pública*. Washington, D.C., Pan American Health Organization.

[246] Parkhurst, P. E. (1975). Generating Meaningful Hypotheses with Aptitude-Treatment Interactions. *AV Communication Review*. 23 (2), 171–184.

[247] Pateman, C. & Grosz, E. (2013). *Feminist Challenges: Social and Political Theory*. London: Routledge.

[248] Patton, M. (1990). *Qualitative Evaluation and Research Methods* (2nd Ed.). Newbury Park, London, New Delhi: Sage Publications.

[249] Paul, J. (2005). *Introduction to the Philosophies of Research and Criticism in Education and the Social Sciences*. Upper Sanddle River, New Jersey: Pearson & Merril Prentice Hall.

[250] Pavlova, M. (2015). Design and Technology Education for Sustainable Future. In K. Stables and S. Keirl. *Environment, Ethics and Cultures* (pp. 87–99). Rotterdam: Sense Publishers.

[251] Pe'er, S., Yavetz, B. & Goldman, D. (2013). Environmental Education for Sustainability as Values Education. Embracing the Social and the Creative: New Scenarios for Teacher Education.

[252] Pedretti, E. (2014). Environmental Education and Science Education: Ideology, Hegemony, Traditional Knowledge, and Alignment. *Revista Brasileira de Pesquisa em Educação em Ciências*, 14 (2), 305–314.

[253] Perakyla, A. (2005). Analyzing taLk and Text. En N. K. Denzin & Y. S. Lincoln (eds.), *Sage Handbook of Qualitative Research* (pp. 869–886). Thousand Oaks, CA: Sage.

[254] Perkins, S. (2014). Geoscience: Fracking Fundamentals. *Nature*, 507, 263–264.

[255] Pichardo, M. C., García, A. B., De la Fuente, J. & Justicia, F. (2007). El Estudio de las Expectativas en la Universidad: Análisis de Trabajos Empíricos y Futuras Líneas de Investigación. *Revista Electrónica de Investigación Educativa*, 9 (1).

[256] Plano, V. & Crewell, J. (2008). *The Mixed Methods Readers*. Los Ángeles: Sage Publications.

[257] Pojman, L., Pojman, P. & McShane, K. (2015). *Environmental Ethics: Readings in Theory and Application*. Toronto: Nelson Education.

[258] Ponce, O. (2011) *Investigación de Métodos Mixtos en Educación: Filosofía y Metodología*. Puerto Rico: Publicaciones Puertorriqueñas.

[259] Ponce, O. (2014). *Investigación Cualitativa en Educación: Teoría, Prácticas y Debates*. Hato Rey, Puerto Rico: Publicaciones Puertorriqueñas.

[260] Ponce, O. (2014). *Investigación Cualitativa en Educación: Teorías, Prácticas y Debates*. San Juan, PR: Publicaciones.

[261] Pooley, J. A. & O'Connor, M. (2000). Environmental Education and Attitudes. Emotions and Beliefs are What is Needed. *Environmental and Behavior*, 32 (5), 711–723.

[262] Potrafke, N. & Ursprung, H. W. (2012). Globalization and Gender Equality in the Course of Development. *European Journal of Political Economy*, 28, 399–413.

[263] Potter, V. R. (1971). *Bioethics: Bridge to the Future*. Englewood Cliffs: Prentice-Hall.

[264] Price, L. (2006). Gender Differences and Similarities in Online Courses: Challenging Stereotypical Views of Women. *Journal of Computer Assisted Learning*, 22 (5), 349–359.

[265] Prings, R. (2000). *Philosophy of Educational Research* (2nd Ed.). London: Continuum.

[266] Punch, K. F. (2005). *Introduction to Social Research: Quantitative and Qualitative Approaches*. Thousand Oaks, CA: Sage.

[267] Rabin C. (1998). Gender and Intimacy in the Treatment of Couples in the 1990s. *Sexual and Marital Therapy*, 13, 179–199.

[268] Rebollo, M. A., García Pérez, R., Piedra, J. & Vega, L. (2011). Gender Culture Assessment in Education: Teachers' Attitudes to Gender Equality. *Revista de Educación*, 521–546.

[269] Regan, T. (1993). *The Case for Animal Rights*. Berkeley: University of California Press.

[270] Rivera, N. (2008). *Analysis of the Expectations and Educative Needs of the Professionals in the Retirement Stage: An Andragogic Alternative*. (Unpublished Doctoral Dissertation). Cupey, PR: School of Education, Inter American University of Puerto Rico.

[271] Robles, J. (2000). The Need for an Ethic of the Land: Living as if Nature Mattered. Deliberation on the Development of Environmental Bioethics. Puerto *Rico Health Sciences Journal*, 19 (3), 289–296.

[272] Rogers, E. M. (1986). *Communication Technology: The New Media in Society*. New York: Free Press.

[273] Rosenbluth, S., Steil, J. & Whitcomb, J. (1998). Marital Equality: What does it Means? *Journal of Family Issues*. 19, 227–244.

[274] Rowe, D. (2007), Education for Sustainable Future, *Science*, 317, 323–324.

[275] Sagor, R. (2005). *The Action Research Guidebook: A Four-Step Process for Educators and Teams*. Thousand Oaks: Corwin.

[276] Saldana, J. (2013). *The Coding Manual for Qualitative Researchers*. Thousand Oaks, CA: Sage.

[277] Salomon, G. (1981). *Communication and Education: Social and Psichological interactions*. Beverly Hills: Sage Publications.

[278] Salomone, R. (2007). Igualdad y Diferencia. La Cuestión de la Equidad de Género en la Educación. *Revista Española de Pedagogía*, 238, 433–446.

[279] Sarmiento, P. J. (2001). Bioética y Medio Ambiente: Introducción a la Problemática Bioético-Ambiental y sus Perspectivas. *Persona y Bioética*, 5, 6–35.

[280] Sarmiento, P. J. (2013). Bioética Ambiental y Ecopedagogía: Una Tarea Pendiente. *Acta Bioethica*, 19 (1), 29–38.

[281] Sauvé, L. (1999). La Educación Ambiental entre la Modernidad y la Posmodernidad: En Busca de un Marco de Referencia Educativo Integrador. *Tópicos en Educación Ambiental*, 1 (2), 7–25.

[282] Scheer, H. (2012). *The Energy Imperative*. Abingdon: Routledge.

[283] Schofield, J. (2007). Increasing generalizability of qualitative research. (Chapter 13. In Hammersley, M. (Ed.). *Educational Research and Evidence-Based Practice*. Los Angeles, London, New Delhi, Singapore: Sage Publications.

[284] Schramm, W. (1954). *Procedures and Effects of Mass Communication*. New York: Basic.

[285] Schramm, W. (1957). *Responsibility in Mass Communication*. New York: Harper & Brothers.

[286] Scott, W. (2015). Education for Sustainable Development (ESD): A Critical Review of Concept, Potential and Risk. In R. Jucker & R. Mathar *Schooling for Sustainable Development in Europe* (pp. 47–70). Dordrecht: Springer.

[287] Scott, W. & Oulton, C. (1998). Environmental Values Education: An Exploration of its Role in the School Curriculum. *Journal of Moral Education*, 27, 209–224.

[288] Seguino, S. (2000). Gender Inequality and Economic Growth: A Cross-Country Analysis, *World Development*, 28 (7), 1211–30.

[289] Sewell, G. (2015). *In the Beginning: And other Essays on Intelligent Design*. Seattle: Discovery Institute.

[290] Shanks, N. & Green, K. (2011). Intelligent Design in Theological Perspective. *Synthese*, 178 (2), 307–330.

[291] Shannon, C. E. & Weaver, W. (1949). *The Mathematical Theory of Communication*. Urbana, IL: University of Illinois.

[292] Silander, C., Haake, U. & Lindberg, L. (2013). The Different Worlds of Academia: A Horizontal Analysis of Gender Equality in Swedish Higher Education. *Higher Education*, 66, 173–188.

[293] Silverman, D. (2005). *Doing Qualitative Research* (2nd). London, Thousand Oaks, New Delhi: Sage Publications.

[294] Singer, P. (1995). *Animal Liberation*. New York: Random House.

[295] Singh, K., Luseno, W. & Haney, E. (2013). Gender Equality and Education: Increasing the Uptake of HIV Testing among Married Women in Kenya, Zambia and Zimbabwe. *Aids Care-Psychological and Socio-Medical Aspects of Aids/Hiv*, 25, 1452–1461.

[296] Sinnes, A. (2012). Three Approaches to Gender Equity in Science Education. *Nordic Studies in Science Education*, 2 (1), 72–83.

[297] Smyth, J. C. (1996). Environmental Values and Education. In J. M. Halstead & T. Taylor (Eds.), *Values in Education and Education in Values* (pp. 54–67). London: The Falmer Press.

[298] Solangi, K. H. et al. (2011). A Review on Global Solar Energy Policy. *Renewable and Sustainable Energy Reviews*, 15 (4), 2149–2163.

[299] Solomon, S. et al. (eds.). (2007). *IPCC Climate Change 2007: The Physical Science Basis. Contribution of Working Group I to the Fourth Assessment Report of the Intergovernmental Panel on Climate Change*. Cambridge & New York: Cambridge University Press.

[300] Sorlin, A., Ohman, A., Ng, N. & Lindholm, L. (2012). Can the Impact of Gender Equality on Health be Measured? A Cross-Sectional Study Comparing Measures Based on Register Data with Individual Survey-Based Data. *Bmc Public Health*, 12.

[301] Sosa, N. M. (1990). *Ética Ecológica*, Madrid: Ediciones Libertarias.

[302] Statistical Package for the Social Sciences (SPSS). IBM SPSS Statistics 21.0. (August, 2012). *IBM Corporation*.

[303] Sterling, S. (2001). *Sustainable Education*. London: Green Books.

[304] Sterling, S. & Huckle, J. (Eds.). (2014). *Education for Sustainability*. London: Routledge.

[305] Stewart, J. F., Choi, J. & Mallery, C. (2010). A Multilevel Analysis of Distance Learning Achievement: Are College Students with Disabilities Making the Grade? *Journal of Rehabilitation*, 76 (2), 27–39.

[306] Strauss, A. & Corbin, J. (1990). *Basics of Qualitative Research: Grounded Theory Procedures and Techniques*. Newbury Park, London, New Delhi: Sage Publications.

[307] Tashkkori, A. & Teddlie, C. (1998). *Mixed Methodology: Combining Qualitative and Quantitative Approaches*. Thounsand Oaks: Sage Publications.

[308] Teddlie, C. & Tashakkori, A. (2009). *Foundations of Mixed Methods Research: Integrating Quantitative and Qualitative Approaches in the Social and Behavioural Sciences*. California: Sage.

[309] Teelken, C. & Deem, R. (2013). All are Equal, but Some are More Equal than Others: Managerialism and Gender Equality in Higher Education in Comparative Perspective. *Comparative Education*, 49, 520–535.

[310] Tester, K. (2015). *Animals and Society: The Humanity of Animal Rights*. New York: Routledge.

[311] Thakran, S. (2015). Education for Sustainable Development. *Educational Quest*, 6 (1), 55–60.

[312] Thomas, G. & Pring, R. (2004). *Evidence-Based Practice in Education*. Maidenhead: Open University Press.

[313] Thompson, P. B. (2015). From Synthetic Bioethics to One Bioethics: A Reply to Critics. *Ethics, Policy & Environment*, 18 (2), 215–224.

[314] Tickell, C. (1995). Education for Sustainability. *Environmental Education*, 50, 5–7.

[315] Trajber, R. & Mochizuki, Y. (2015). Climate Change Education for Sustainability in Brazil: A Status Report. *Journal of Education for Sustainable Development*, 9 (1), 44–61.

[316] Tyer-Viola, L. A. & Cesario, S. K. (2010). Addressing Poverty, Education, and Gender Equality to Improve the Health of Women Worldwide. *Jognn-Journal of Obstetric Gynecologic and Neonatal Nursing*, 39, 580–589.

[317] Tytler, R. (2011). Socio-Scientific Issues, Sustainability and Science Education. *Research in Science Education*, 42 (1), 155–163.

[318] UNESCO (1978). *Final Report, Intergovernmental Conference on Environmental Education*. Tbilisi: Author.

[319] UNESCO (1984). *La Educación en Materia de Comunicación*. Paris: Author.

[320] UNESCO (1997). *Education for Sustainable Future: A Transdisciplinary Vision for Concerted Action.* Paris: UNESCO.

[321] UNESCO (2010). *Climate Change Education for Sustainable Development.* Paris: Author.

[322] UNFRA (2012). *Estado de la Población Mundial 212.* New York: Fondo de Población de las Naciones Unidas.

[323] Unterhalter, E. (2005). Global Inequality, Capabilities, Social Justice: The Millennium Development Goal for Gender Equality in Education. *International Journal of Educational Development,* 25 (2), 111–122.

[324] Van Manen, M. (1990). *Researching Lived Experience: Human Science For An Action Sensitive Pedagogy.* London, Ontario, Canada.

[325] Van Willigen, M. & Drentea, P. (2001). Benefits of Equitable Relationships: The Impact of Sense of Fairness, Household Division of Labor, and Decision Making Power on Perceived Social Support. *Sex Roles,* 44, 571–597.

[326] Velasco, I. (2007). Hacia una Educación superior con Equidad de Género. *Inventio,* 5, 43–48.

[327] Venkataraman, B. (2009). Education for Sustainable Development. Environment: *Science and Policy for Sustainable Development,* 51 (2), 8–10.

[328] Walliman, N. (2011). *Research Methods: The Basics.* London: Routledge.

[329] Waters, C. K. (2007). The Nature and Context of Exploratory Experimentation: An Introduction to Three Case Studies of Exploratory Research. *History & Philosophy of the Life Science,* 29 (3), 275–284.

[330] Weaver-Hightower, M. B. & Skelton, C. (2013). Gender and Education. In C. Skelton (ed.). Leaders in Gender and Education (pp. 1–13). London: Sense Publishers.

[331] Weiler, K. (2001). *Feminist Engagements.* New York: Routledge.

[332] Wellington, J. (2015). *Educational Research: Contemporary Issues and Practical Approaches.* New York, NY: Bloomsbury Publishing.

[333] White, S. S. (2005). Environmental Education in Graduate Professional Degrees: The Case of Urban Planning. *Journal of Environmental Education,* 36 (3), 31–38.

[334] Wolcott, H. F. (1994). *Transforming Qualitative Data.* Thousand Oaks, CA: Sage.

[335] Woods, P. (1996). *Researching the Art of Teaching. Ethnography for Educational Use.* Routledge. London and New York: State University Of New York Press and The University of Western Ontario.

[336] World Health Organization (2005), *Multi-Country Study on Women's Health and Domestic Violence Against Women*. Geneva: World Health Organization.

[337] Wrigley, J. (Ed.). (2003). *Education and Gender Equality*. London: Routledge. Aliciardi, M. B. (2009). ¿Bioética Ambiental? *Revista Latinoamericana de Bioética*, 9 (1), 8–27.

[338] Yin, R. (1994). *Case Study Research: Designs and Methods* (2nd Ed.). Applied Social Research. Methods Series, Volume 5. Thousand Oaks: Sage Publications.

[339] Zenelaj, E. (2013). Education for Sustainable Development. *European Journal of Sustainable Development*, 2 (4), 227–232.

Index

About the Editor

José Gómez Galán is currently *Research Professor* at the Metropolitan University (AGMUS, Puerto Rico – United States) and *Professor* of Theory and History of Education at the University of Extremadura (Spain). *Visiting Researcher* and *Professor* at several international universities: University of Oxford (UK), University of Minnesota (USA), Università degli Studi La Sapienza of Rome (Italy), several Latin American universities, etc. He is Doctor (Ph.D.) in Philosophy and Education, Doctor (Ph.D.) in Geography and History, and holds Masters/DEA degrees and others academics degrees in various international universities. His main research in a multidisciplinary context, focus on the scientific fields of Education, Social and Natural Sciences in news lines of interdisciplinary dialogue. He is a director and member of various research groups in different academic centers on national and international scale. He is the author of numerous scientific publications, both nationally and internationally (books, book chapters, articles in scientific journals, etc.), in many different fields of knowledge. He has participated in distinguished congresses, seminars and international symposiums, and is director or partner in various research projects in prestigious centers and universities in Europe and the United States. He is currently a member of committees (editorial, scientific and advisory) and/or referee of important international journals included in the Journal Citation Reports (JCR). He has received several major awards for teaching and research, and was awarded the National Educational Research Award (Spain).